高砷矿区食物链中砷的生物富集特征与转化行为

杨 芬 编著

电子工业出版社
Publishing House of Electronics Industry
北京·BEIJING

未经许可，不得以任何方式复制或抄袭本书之部分或全部内容。
版权所有，侵权必究。

图书在版编目（CIP）数据

高砷矿区食物链中砷的生物富集特征与转化行为 / 杨芬编著. -- 北京 : 电子工业出版社, 2024. 10.
ISBN 978-7-121-49056-9

Ⅰ. X5

中国国家版本馆 CIP 数据核字第 20249GH178 号

责任编辑：缪晓红
印　　刷：涿州市般润文化传播有限公司
装　　订：涿州市般润文化传播有限公司
出版发行：电子工业出版社
　　　　　北京市海淀区万寿路 173 信箱　邮编 100036
开　　本：720×1 000　1/16　印张：16　字数：256 千字
版　　次：2024 年 10 月第 1 版
印　　次：2024 年 10 月第 1 次印刷
定　　价：98.00 元

凡所购买电子工业出版社图书有缺损问题，请向购买书店调换。若书店售缺，请与本社发行部联系，联系及邮购电话：(010) 88254888，88258888。
质量投诉请发邮件至 zlts@phei.com.cn，盗版侵权举报请发邮件至 dbqq@phei.com.cn。
本书咨询联系方式：010-88254760。

本书的出版得到国家自然科学基金项目"食物来源影响水生生物富集与转化砷的差异性及其机制（41907325）""矿区食物链中砷的生物富集与转化特征（41571470）"的支持。同时，本书的研究结论是以上两个项目中部分工作的总结。

前　言

　　采矿和冶炼活动所造成的环境问题已经成为全球性问题，引起了越来越多学者的关注。重金属在环境中进行迁移时，一旦进入食物链，就可能由于生物累积作用在生物体内蓄积。砷的富集能通过食物链进入人体，可能会产生健康风险。砷对生物体的毒性与砷的浓度、形态、生物有效性及摄取方式有关。一般来说，无机砷的毒性高于有机砷，三价砷的毒性高于五价砷。在砷污染区，人体砷暴露途径主要为饮水、食物等，为降低砷对人体的潜在健康风险，迫切需要弄清高砷污染区砷的分布特征及其在环境中的生物地球化学行为。但迄今为止，仅有零星的研究关注砷矿区砷在土壤、水体中的分布，以及部分生物对砷的富集特征。关于砷污染区砷的生物迁移累积和完整食物链中砷的富集效应、砷形态特征等研究报道还不多见。因此，研究不同营养级生物对砷的富集和转化特征成为砷污染环境健康效应及其修复研究的热点之一。

　　砷是一种有毒有害并致癌的元素，在环境中普遍存在。湖南常德地区的石门雄黄矿是亚洲最大的雄黄矿，其开采与冶炼已历经 1500 多年。多年冶炼活动的进行，加之含砷废水、废气及废渣的不当处置等，导致大量的砷遗留在土壤和水体等环境介质中。本书重点研究石门雄黄矿典型食物链，尤其是在淡水水生生物和陆地生物中砷的生物富集与转化行为之间的关系。具体内容包括：首先，在不同程度的砷污染区采集不同类型的生物和环境样本，分析砷在食物链各营养级的生物和环境中的含量和形态特征，全面揭示石门雄黄矿食物链中砷的生物富集与转化行为；其次，以胃含物分析结果为基础，应用 MixSIAR 模型，采用碳氮稳定同位素定量分析不同食性水生生物的潜在食物来源组成及其贡献率，开展不同食性水生生物及其食物来源中砷含量及形态特征综合对比分析，揭示食物来源对水生生物富集和转化砷的影响机制。

本书主要由杨芬撰写，中国科学院地理科学与资源研究所韦朝阳研究员为书稿的撰写提供了宝贵意见，也为本书的校对工作付出了辛勤劳动。本书出版得到了国家自然科学基金项目"食物来源影响水生生物富集与转化砷的差异性及其机制（41907325）""矿区食物链中砷的生物富集与转化特征（41571470）"的支持。同时，本书的研究结论是以上两个项目中部分工作的总结。

本书通过对石门雄黄矿不同程度砷污染区中淡水生态系统和陆地生态系统对砷的生物富集和转化行为研究，进一步应用碳氮稳定同位素和胃含物相结合的技术手段分析不同食性水生生物的潜在食物来源及其组成变化，阐明不同食物来源对砷的富集和转化砷的可能贡献和影响机制，对我国代表性地区生态系统砷污染防治及其生物地球化学行为等方面提供一些新认识。

本书可作为环境科学、环境工程及相关领域专家学者的参考读物，也可为广大环境保护工作者提供借鉴。

目 录

第1章 绪论 ··· 001

1.1 砷的富集与转化研究的背景与意义 ·· 003

1.2 砷的来源、富集与转化和研究方法的进展 ·· 006

    1.2.1 砷的来源及其毒性研究进展 ·· 006

    1.2.2 生物体中的砷研究进展 ·· 010

    1.2.3 食物链（网）中砷的富集与转化效应 ··· 015

    1.2.4 稳定同位素确定营养级方法概述 ··· 018

    1.2.5 砷形态分析方法研究进展 ··· 020

1.3 砷矿区食物链中砷的生物富集与转化研究思路 ···································· 026

    1.3.1 砷矿区食物链中的生物富集与转化研究目的 ······························· 026

    1.3.2 砷矿区食物链中的生物富集与转化研究内容 ······························· 027

参考文献 ··· 028

第2章 砷矿区概况及环境和生物样品砷形态和砷含量分析 ····················· 053

2.1 石门雄黄矿概述 ··· 055

    2.1.1 石门雄黄矿自然地理位置概况 ··· 055

    2.1.2 石门雄黄矿地质及开采历史概况 ··· 056

2.2 石门雄黄矿环境及生物样品实验方法 ·· 059

    2.2.1 环境和生物样品采集及预处理 ··· 059

2.2.2　石门雄黄矿环境和生物样品分析 ································· 061
　　2.2.3　石门雄黄矿环境和生物样品总砷量测定方法 ············· 064
　　2.2.4　石门雄黄矿环境和生物样品砷形态分析方法 ············· 066
　　2.2.5　石门雄黄矿环境和生物样品中其他重金属含量测定方法 ····· 069
2.3　石门雄黄矿环境和生物样品总砷和砷形态质量控制 ············· 071
　　2.3.1　石门雄黄矿环境和生物样品总砷质量控制 ················· 071
　　2.3.2　石门雄黄矿环境和生物样品砷形态质量控制 ············· 072
2.4　本章小结 ··· 074
参考文献 ··· 075

第3章　砷矿区环境介质中砷的分布特征 ································ 079
3.1　砷矿区环境介质中砷的分布的研究意义 ····························· 081
3.2　砷矿区环境介质采集、总砷及砷形态分析方法 ··················· 083
3.3　砷矿区环境介质中砷的分布特征 ······································ 084
　　3.3.1　水体中砷的分布特征 ·· 084
　　3.3.2　沉积物中砷的分布特征 ··· 090
　　3.3.3　土壤中砷的分布特征 ·· 094
3.4　本章小结 ··· 098
参考文献 ··· 099

第4章　陆地环境中砷的生物富集与转化特征 ······················· 105
4.1　陆地环境中砷的生物富集与转化研究意义 ·························· 107
4.2　砷矿区陆地环境生物样品采集、总砷及砷形态分析方法 ········ 110

4.3 陆地环境中砷的生物富集与转化特征·· 114
    4.3.1 土壤—植物—凋落物—土壤动物系统中砷的富集特征········ 114
    4.3.2 土壤—植物—凋落物—土壤动物系统中砷的形态特征········ 119
    4.3.3 鸟类中砷的生物富集与转化特征·· 124
    4.3.4 家禽母鸡中砷的生物富集与转化特征·· 129
    4.3.5 陆地生态系统中砷的转化·· 134
4.4 本章小结··· 136
参考文献··· 137

第5章 水生环境中砷的生物富集与转化·· 151
5.1 水生环境中砷的生物富集与转化研究意义······································· 153
5.2 砷矿区水生环境中生物样品采集、总砷及砷形态分析方法········· 156
5.3 水生环境中砷的生物富集与转化··· 159
    5.3.1 浮游生物多样性与丰度分布·· 159
    5.3.2 水生生物对砷的富集特征及其影响因素·································· 161
    5.3.3 不同营养级水生生物对砷的富集效应······································ 169
    5.3.4 水生生物中的砷形态及其转化·· 175
    5.3.5 石门雄黄矿食用水生生物的健康风险评估······························ 188
5.4 本章小结··· 191
参考文献··· 192

第6章 食物来源影响水生生物富集和转化重金属的差异性········ 203
6.1 食物来源影响水生生物富集和转化重金属的研究意义················· 205

6.2 砷矿区水生环境生物样品采集、总砷及砷形态分析方法 ………… 211
6.3 水生生物及其食物来源对重金属的富集和砷形态分布 ………… 215
 6.3.1 水生生物胃内容物和碳氮同位素分析 ………………… 215
 6.3.2 水生生物及其食物来源的重金属含量 ………………… 218
 6.3.3 水生生物及其食物来源的砷形态分布特征 …………… 222
6.4 水生生物及其食物来源对重金属的富集和砷的转化 …………… 226
 6.4.1 水生生物食物来源组成特征 …………………………… 226
 6.4.2 水生生物及其食物来源对重金属的生物富集和营养传递 … 228
 6.4.3 水生生物及其食物来源对砷的转化特征 ……………… 229
6.5 本章小结 …………………………………………………………… 232
参考文献 ………………………………………………………………… 233

扫码可查看本书部分彩图

第 1 章

绪 论

1.1 砷的富集与转化研究的背景与意义

我国将雄黄（As_2S_2）、雌黄（As_2S_3）等砷化物用于食用、制药具有悠久的历史（李慧和张立实，2000）。自1250年马格耐斯分离出砷之后，砷被广泛应用于医药、农业、畜牧业、电子、工业及冶金业等领域。砷（Arsenic，As）是地表环境中普遍存在的微量元素，微量砷在日常饮食、饮水、空气和土壤中都有发现（Mandal and Suzuki，2002）。

砷及其化合物对动植物有广泛的致毒性，被世界卫生组织（WHO）下属的国际癌症研究所（IARC）、美国环境卫生科学研究院（NIEHS）、美国环保局（USEPA）等诸多权威机构和组织认定为人类一类致癌物（Goering et al.，1999；Ng et al.，2003）。人体摄入砷可能导致皮肤、肺、肝、肾、膀胱等器官的病变，乃至诱发癌症（Ng et al.，2003）。饮水砷中毒在多个国家和地区持续发生，孟加拉国、印度、中国、阿根廷、智利、匈牙利、墨西哥、罗马尼亚、越南及美国等都报道过人群急慢性饮水砷中毒事件（Nikolaidis et al.，2004）。随着人们对砷的毒性的进一步认识和环境健康风险意识的提高，砷污染已成为全球关切的重大环境与健康问题之一。

中国矿产资源丰富，已经发现171种矿产资源，已探明储量的矿产有158种（蒲含勇和张应红，2001）。中国是世界上砷污染较严重的国家之一，矿业活动是导致砷污染的重要原因之一。

在矿产采冶过程中，采出的砷有70%弃留于尾矿中，矿业活动释放的砷可通过土壤及食物链威胁到人体健康（肖细元 等，2008）。砷矿石在燃烧熔炼时，砷易变成蒸气，在空气中迅速被氧化成三氧化二砷（As_2O_3），凝结成固体粒子污染大气、土壤和水体等。冶炼砷时所产生的废气，其污染半径可达50 km，从而危及周围居民的健康（徐红宁和许嘉琳，1996）。

湖南常德地区石门县雄黄矿是亚洲最大的雄黄产地，已有1500多年的开采历史。该矿区自1951年到2012年，确诊的砷慢性中毒者1000多人，有近400人死于砷中毒诱发的各种癌症，其中肺癌近300人（黄芳，2014）。尽管2001年该矿正式闭矿，但是多年的矿产采冶活动，导致大量的砷进入水、土壤等环境中，造成了长期严重的砷污染。Tang等人（2016）研究发现，石门县雄黄矿中心区域土壤中的砷含量最高可达到5240 mg/kg，Zhu等人（2015）对尾矿和冶炼区水体中的砷含量进行检测，发现砷含量高达40.10 mg/L。

采冶活动所造成的环境问题已经成为全球性问题，引起了越来越多学者的关注。重金属在环境中进行迁移时，一旦进入食物链，就可能由于生物累积作用在生物体内蓄积。砷的富集能通过食物链进入人体，可能会导致健康风险。在环境中，砷形态主要为无机砷，无机砷进入生物体后，经过氧化还原和甲基化过程，产生多种甲基砷（Hughes，2002）。砷对生物体的毒性与砷的浓度、形态、生物有效性及摄取方式有关。一般来说，无机砷的毒性高于有机砷，三价砷的毒性高于五价砷（Azizur Rahman and Hasegawa，2012）。

在砷污染区，人体砷暴露途径主要为饮水、食物等，为降低砷对人体的潜在健康风险，我们迫切需要弄清楚高砷污染区砷的分布特征及其在环

第1章 绪 论

境中的生物地球化学行为。但迄今为止,仅有零星的研究关注砷矿区的砷在土壤、水体中的分布,以及部分生物对砷的富集特征。关于砷污染区砷的生物迁移累积和完整食物链中砷的富集效应、砷形态特征等研究报道还不多见。因此,不同营养级生物对砷的富集和转化特征成为砷污染环境健康效应及其修复研究的热点之一。

综上所述,砷对环境与健康的影响已日趋严重,但目前对于砷的形态转化、生物富集特征和机理等方面的认识严重不足,急需开展更多的相关研究。

本研究以石门县雄黄矿为研究对象,通过对该矿区及周边地区陆生和水生食物链中不同营养级生物对砷的富集和转化特征的分析,对砷在砷矿区环境中的时空分布和在生物中的传递作用与生物致毒效应建立一些新的认识,对砷在生物中的生物地球化学行为形成较为系统的理解,为深入认识环境中砷的迁移、转化、富集及毒性效应提供基础理论支撑,为砷矿区砷污染的防治及健康风险应对等方面提供科学依据和建设性意见。

1.2 砷的来源、富集与转化和研究方法的进展

1.2.1 砷的来源及其毒性研究进展

1.2.1.1 砷的来源研究进展

随着工农业的迅猛发展，采矿冶炼、磷肥和杀虫剂施用、能源燃料制造等人类活动，使得大量的重金属被开采使用。这些重金属在开采、运输、加工、使用、废弃的过程中不断被扩散到环境中，包括大气、水体、土壤、生物体等（黄益宗 等，2013）。相较于其他介质，土壤中的重金属一般具有分布范围广、残留时间久、在多形态间转化、不易降解、难于治理等特性。土壤中某些过量元素成分不但影响农业生产，而且还能通过食物链最终对人类健康造成威胁。

金属矿山在开采过程中，产生大量含有重金属的尾矿和酸性矿山废水。在冶炼过程中，会产生大量的含尘气体和矿渣，这些含尘气体和矿渣携带

着大量的重金属进入土壤和水环境中，从而直接或间接造成环境重金属污染（和莉莉 等，2008；仇荣亮等，2009）。因此，矿山或者冶炼厂及周围土壤中重金属污染的报道多有发生。

砷在地壳中的平均丰度为 1.8 mg/kg。环境中的砷有两个来源：自然来源和人为来源。自然来源包括火山、热泉活动和岩石风化。砷存在于约 200 多种不同的含砷矿物和化合物之中，其中约 60%是砷酸盐，20%是硫化物和硫化矿物，其余 20%包括砷化物、亚砷酸盐、氧化物、硅酸盐及单质砷（Mandal and Suzuki，2002）。人为来源包括采矿、选矿、金属冶炼、化石燃料燃烧、木材燃烧和含砷化学品使用（除草剂、杀虫剂、干燥剂、防腐剂、肥料、染料、电子工业原料、制药及军事弹药等）等导致的砷排放（Matschullat，2000；Madigan et al.，2005）。据估算，全球范围内每年人为活动向大气、水体和土壤中排放的砷总量分别达 $12\times10^6 \sim 25.63\times10^6$ kg、$12\times10^6 \sim 70\times10^6$ kg 和 $52\times10^6 \sim 112\times10^6$ kg（Nriagu and Pacynat，1988）。

地表水和地下水中砷的含量为 $1 \sim 10\ \mu g/L$；河口水中砷的含量一般低于 $4\ \mu g/L$；而海水中砷的含量较低且波动相对较小，一般低于 $1.5\ \mu g/L$（Sharma and Sohn，2009）。据估算，全球海洋水体中总砷量达 $2.38\times10^9 \sim 5.15\times10^9$ t，河水中总砷量为 $5.4\times10^4 \sim 6.1\times10^4$ t；每年经海底火山作用和大气沉降作用进入海洋的砷含量分别为 4870 t 和 $4300 \sim 8200$ t，经海底沉积作用和海洋蒸发释放的砷含量则为 84600 t 和 27 t，由于人为活动进入河水中的砷含量则为 12500 t（Matschullat，2000）。

地下水中砷的来源主要有含砷矿床的开采、含砷农药的使用、农业灌溉、木材保存和含砷废水的排放等人为活动，以及地下岩层中含砷矿物/颗粒物中砷溶解/解吸等自然过程（Mandal and Suzuki，2002）。地表水环境中砷的来源主要包括：地下水灌溉、采矿、工业"三废"排放、矿物岩石中砷的自然释放、大气中砷的沉降、土壤侵蚀过程中砷的淋溶释放等

（Nriagu and Pacynat，1988；Matschullat，2000）。矿山开采是环境中砷的重要来源，某些矿区水体中的砷含量可达 5000 μg/L（Smedley and Kinniburgh，2002）。大气中的砷主要来自铜矿冶炼和燃煤等人为活动，再通过干湿沉降进入土壤和水体，从而增加环境中砷的浓度（Smedley and Kinniburgh，2002）。每年从大气中释放到水体中的砷约为 $3.6\times10^6 \sim 7.7\times10^6$ kg（Matschullat，2000）。土壤中的砷主要来自岩石层风化、大气砷沉降，以及矿物开采、农药化肥施用、工业废水灌溉等工农业活动，通过侵蚀和淋溶作用进入水体。同时土壤中的微生物活动还能加速砷的淋溶作用，使砷快速进入水体之中（Turpeinen et al.，1999）。

1.2.1.2 砷的毒性研究进展

砷是一种有毒并致癌的化学元素，普遍存在于各种环境介质和生物体中（Mandal and Suzuki，2002）。淡水系统中的砷以 4 种价态（-III，0，III，V）存在，无机砷主要包括 As(III) 和 As(V)，有机砷主要包括一甲基砷酸（MMA）、二甲基砷酸（DMA）、三甲基砷酸（TMA），而在海洋生物中则主要为砷甜菜碱（AsB）和砷胆碱（AsC）（见表 1.1）（Sharma and Sohn，2009）。

表 1.1 淡水系统中主要的砷化合物及其化学结构式（Leybourne et al.，2014）

Chemical forms of arsenic found in freshwater systems

名　　称	化　学　式	缩　　写
无机砷：		
Arsenite 亚砷酸	H_3AsO_3	As(III)
Arsenate 砷酸	H_3AsO_4	As(V)
甲基砷（有机砷）：		
Monomethylarsonous acid 一甲基砷酸	$CH_3As(OH)_2$	MMA(III)
Monomethylarsonic acid 一甲基砷酸	$CH_3AsO(OH)_2$	MMA(V)
Dimethylarsinous acid 二甲基砷酸	$(CH_3)_2As(OH)$	DMA(III)

续表

名　称	化　学　式	缩　写
Dimethylarsinic acid 二甲基砷酸	$(CH_3)_2AsO(OH)$	DMA(V)
Trimethylarsine acid 三甲基砷酸	$(CH_3)_3As$	TMA
Trimethylarsinic oxide 三甲基砷氧化物	$(CH_3)_3AsO$	TMAO
Tetramethylarsonium ion 四甲基砷离子	$(CH_3)_4As^+$	TeMAs$^+$
其他有机砷：		
Arsenocholine 砷胆碱	$(CH_3)_3As^+CH_2CH_2OH$	AsC
Arsenobetaine 砷甜菜碱	$(CH_3)_3As^+CH_2COO^-$	AsB
Arsenosugars 砷糖	$(CH_3)_2AsOCH_2C_4H_4O\text{-}R$[①]	AsS
Thioarsenates（1-，2-，3-）硫代砷酸盐		TAs(V)

注：①R 为 OH、$OPO_3CH_2CH(OH)CH_2OH$、SO_3H、OSO_3H 等。

环境中的砷通过皮肤、呼吸道、消化道等进入人体，在肝、肾、肺、脾、子宫、胎盘、骨骼、肌肉等器官和组织中缓慢积累，通过代谢作用在毛发和指甲等末端组织上积聚（Hughes，2002；Ng et al.，2003；Liu et al.，2017）。砷在人体中积累长达几年甚至几十年之后，会出现一系列的砷中毒现象，包括皮肤损伤，以及呼吸系统、消化系统、循环系统、神经系统损害和器官变异，最终导致各种癌症，如皮肤癌、肺癌、膀胱癌、胃癌、肝癌和肾癌等（Goering et al.，1999；Matschullat，2000；Mandal and Suzuki，2002；Sharma and Sohn，2009；Hughes et al.，2011）。砷的毒性极高，当人体毛发砷含量达到 1 μg/g 以上，指甲砷含量为 20～130 μg/g，或每天尿液中砷含量超过 100 μg/L 时，极易引起砷中毒（Mandal and Suzuki，2002）。

进入生物体内的砷形态主要为无机砷，无机砷在体内谷胱甘肽（GSH）及 S-腺苷蛋氨酸（SAM）的辅助作用下代谢，过程为 As(V)→As(III)→MMA(V)→MMA(III)→DMA(V)→DMA(III)→TMAO（Hughes，2002）。

大量室内研究发现，人体摄入的 As(III)和 As(V)会被广泛地甲基化，DMA(V)为人体尿液排出的主要甲基砷化物，日常人体尿液中 MMA(V)也占有相当的比例，而 TMAO 作为人体最终的甲基化产物在尿液中含量很少，MMA(Ⅲ)和 DMA(Ⅲ)作为砷甲基化过程的中间物已在长期饮水砷暴露人体的尿液中被检测出来（Aposhian et al.，2000）。

砷对生物体的毒性与砷的剂量、形态、生物有效性及被摄取的方式有关。一般来说，无机砷的毒性高于有机砷，三价砷的毒性高于五价砷（Azizur Rahman and Hasegawa，2012）。

目前对砷的致病机理的认识尚不成熟，无机砷的甲基化被认为是人体和其他生物体砷解毒的重要机制。但近年来有研究发现，砷甲基化的中间产物 MMA(Ⅲ)与 DMA(Ⅲ)可与蛋白质结合，抑制和破坏生物体中的酶，因此可能导致细胞损伤乃至 DNA 改变，而具强致癌性（Shen et al.，2013）。砷化合物的毒性排序大致为 DMA(Ⅲ) > MMA(Ⅲ) > As(Ⅲ) > As(Ⅴ) > DMA(Ⅴ) > MMA(Ⅴ) > TMAO（Hughes，2002；Mandal and Suzuki，2002）。

Styblo 等人（2000）在对老鼠和人体肝细胞的实验中发现，DMA(Ⅲ)和 MMA(Ⅲ)的毒性要远远高于无机的 As(Ⅲ)。Aposhian 等人（2000）第一次从人体尿液中检测到 MMA(Ⅲ)，并发现只有当人体暴露于高砷环境（>161 μg/L）时，MMA(Ⅲ)才会出现。有报道称，在静脉注射亚砷酸盐的老鼠胆汁中也检测到了 MMA(Ⅲ)和 DMA(Ⅲ)（Gregus et al.，2000）。但 MMA(Ⅲ)和 DMA(Ⅲ)作为砷在生物体中代谢的中间产物，进入体外环境中极易转化为Ⅴ价的甲基砷，难以分离和准确测定（Sohrin et al.，1997；Petrick et al.，2000）。在环境中，如在日本最大的湖泊琵琶湖中，也检测到了 DMA(Ⅲ)和 MMA(Ⅲ)，但其浓度均低于 6×10^{-10} mol/L（Sohrin et al.，1997）。目前，我们对 DMA(Ⅲ)和 MMA(Ⅲ)在环境—生物系统中的吸收、转化和毒性效应还很不清楚。

1.2.2 生物体中的砷研究进展

砷广泛存在于地表各种环境介质与生物体中（Mandal and Suzuki，2002；

Zhang et al., 2013)。砷对生态系统的影响是长期的,并可在食物链中富集。人类活动排放到环境中的砷主要是无机态的砷化合物,包括 As(Ⅲ)和 As(Ⅴ)。砷进入环境后,可被植物和动物吸收。例如,植物可吸收土壤中的砷,土壤动物可吸收土壤水分中的砷,河流中的无脊椎动物可以直接吸收水中的砷,还可通过捕食吸收被捕食者体内的砷(Bundschuh et al., 2012; Button et al., 2012)。

1.2.2.1 陆生生物中的砷研究进展

研究表明,全球土壤中砷的背景含量为 5~6 mg/kg,泥炭土和沼泽土中的砷含量为 13 mg/kg,矿区土壤中砷含量相对较高,甚至可达到 23800 mg/kg(Alvarez et al., 2006; Grace and Macfarlane, 2016)。在天然条件下,土壤中可溶性砷仅占很少比例,主要为无机态的 As(Ⅲ)和 As(Ⅴ)(Alvarez et al., 2006; Chen et al., 2008; Button et al., 2011, 2012; Yao et al., 2016; Tang et al., 2016; Grace and Macfarlane, 2016)。

当环境条件发生变化时,如 pH 值、氧化还原电位、有机质、物质组成、温度和气候及微生物作用等都会引起砷在土壤中的迁移与转化(杨芬等,2015)。值得注意的是,残留在土壤中的砷很难被除去,长期积累在土壤中的砷很容易通过农作物等转移到人体及其他生物体中,从而造成潜在危害。因此,土壤砷污染已成为引人注目的环境问题。

一般情况下,土壤中的砷在淋洗作用下可以进入土壤的深层部位,或者通过植物根部进入植物。不同植物对土壤中的砷有不同程度的富集作用。自 Ma 等人(2001)和陈同斌等人(2002)发现蜈蚣草(*Pteris vittata*)能超富集砷以来,关于砷超富集植物的研究及砷污染植物的修复研究无论从理论方面还是实际应用方面已日渐深入。

迄今为止,国际上已发现 10 余种砷超富集植物并对其富集砷的机理开展了很多研究工作,如蜈蚣草、大叶井口边草等(韦朝阳 等,2002;

Lombi et al., 2002; Zhang et al., 2002; Cao et al., 2004; Huang et al., 2008; Zhao et al., 2009; Jeong et al., 2014; Wan et al., 2017）。但是关于全面查清本土植物对砷的富集特征的研究仅有零星报道（Mir et al., 2007; Ruiz-Chancho et al., 2008; 李莲芳 等, 2010; Otones et al., 2011; Bergqvist and Greger, 2012）。

大多数研究表明，陆地植物中的砷形态主要为As(V)、As(III)、MMA和DMA（Quaghebeur et al., 2003; Mir et al., 2007; Chen et al., 2008; Jedynak et al., 2009; Bergqvist and Greger, 2012; Larios et al., 2012; Wei et al., 2015）。

另外也有研究报道，除以上4种砷形态外，陆地植物中还含有少量的TMAO、Tetra、AsB、AsC等（Meharg and Hartley-Whitaker, 2002; Ruiz-Chancho et al., 2008; Zheng and Hintelmann, 2009）。但是到目前为止，关于植物体内有机砷的来源仍不清楚。Lomax 等人（2012）认为植物本身并没有将无机砷转化为有机砷的能力，而是直接吸收微生物产生的有机砷。

相对来说，关于陆生动物中砷的研究较少。蚯蚓作为一种重要的土壤动物，在重金属的迁移和循环中具有重要作用（van Gestel et al., 2009）。它能够在被重金属污染的土壤中生存，并富集一定量的重金属（Dada et al., 2013）。蚯蚓中砷的来源主要为土壤有机质，而凋落物是土壤有机质的一个重要来源（Langdon et al., 2003b; Button et al., 2011）。大多数研究表明，随着土壤中砷浓度的增加，蚯蚓体内砷含量随之增加（Button et al., 2009; García-Gómez et al., 2014; Romero-Freire et al., 2015）。

一般认为，AsB主要出现在海洋生物中，在陆地生物中很少存在。研究发现，蚯蚓中的砷形态主要为As(V)和As(III)，并含少量的AsB（Langdon et al., 2003a; Button et al., 2009, 2012）。另外，研究发现，AsB作为主要的砷形态存在于很多鸟类中（Kubota et al., 2003; Koch et al., 2005）。目前，关于陆生生物中AsB的形成机理并不明晰（Moriarty et al., 2009），急需进一步研究和探讨。

总的来说，在陆地生态系统中，土壤中的砷通过根系进入植物中，最终以凋落物形式分解释放、重新回到土壤中，而土壤中的动物可以吸收土壤和凋落物中的砷。植物和土壤动物位于陆地生态系统食物链的底部，是食物链顶端高营养级生物（如鸟类、鸡等）的潜在食物来源，因此对土壤—植物—凋落物—土壤动物—高营养级生物系统中的砷循环的研究具有重要意义。然而，目前的研究大多仅仅关注土壤—植物系统，或土壤—土壤动物系统等，对陆地生态系统完整生物链尚无全面的研究。

1.2.2.2 水生生物中的砷研究进展

砷主要通过食物和饮水进入水生动物体内，摄入富集砷的水产品会对人体健康带来潜在的危害。近年来由于水体砷污染的频繁报道，水生生物中的砷研究已引起人们的高度关注（张楠 等，2013）。一般来说，水环境中的砷形态主要为无机砷 As(III)和 As(V)，有机砷以 DMA(V)和 MMA(V)为主，有机砷占比较低（Madigan et al.，2005；Ronkart et al.，2007；Barringer et al.，2011；Azizur Rahman and Hasegawa，2012；Caumette et al.，2012）；而 DMA(V)含量又可达到总甲基砷含量的 64%~99%。此外，还存在 DMA(III)和 MMA(III)，但含量比 DMA(V)和 MMA(V)低 1~2 个数量级（Azizur Rahman and Hasegawa，2012）。

到目前为止，我们对各种砷化物在海水中的分布及生物体中的含量与形态特征已有比较全面的认识。海水中的砷主要吸附在悬浮颗粒物上，而溶解态砷含量极低，主要以 As(III)和 As(V)形态存在（Mandal and Suzuki，2002）。海洋生物中的砷含量较高，一般为 5~50 mg/kg（Mandal and Suzuki，2002）。海洋鱼类、甲壳类动物和软体动物中砷化物主要以砷甜菜碱（AsB）形态存在；在海藻中则以砷糖化合物为主要存在形态，而无机砷含量极少（Ciardullo et al.，2010）。海洋生物对砷一般具有很强的富集能力，特别是海洋中的介贝类，砷的富集系数可达 3300（Mandal and Suzuki，2002）。

研究表明，在淡水湖泊中，浮游生物中的砷含量一般较高，浮游植物中的砷以无机砷存在，As(Ⅴ)占比大于92%，而浮游动物中除无机砷外，还含有少量的甲基砷、砷糖及AsB（Caumette et al.，2011，2012）。Rubio-Franchini等人（2016）研究发现，Dula河流域浮游动物对砷的富集系数为250~11035。不同藻对砷的吸收差异也很大，如褐藻中的砷含量比绿藻和红藻中的砷含量要高10倍左右（Levy et al.，2005）。水生植物对砷的富集及形态特征与陆地植物基本一致。Singh等人（2016）研究发现，当土壤中的砷含量为46.2~117 mg/kg时，水生植物对砷的富集量为21.2~735 mg/kg。Bergqvist和Greger（2012）研究发现，在水的沉积物中砷含量为6.68~100200 mg/kg时，水生植物中的砷含量变化范围较广，为0.22~1623 mg/kg。一般来说，潜水植物中的砷含量大于浮游植物（Zhang et al.，2013）。陈国梁（2014）对5种沉水植物加砷水培的实验结果表明，对砷的富集能力上，苦草>黑藻>菹草>狐尾藻>金鱼藻，它们对砷的富集系数都超过220。

水生植物对砷的富集量随着环境砷含量的增加而有所增加（Kim et al.，2009）。水生植物的砷形态主要为无机砷，还含有少量的甲基砷（Zheng et al.，2003；Bergqvist and Greger，2012）。一般认为，水生植物主要通过磷通道吸收As(Ⅴ)（Tripathi et al.，2007；Zhao et al.，2013）。但是也有研究认为，根系表面的物理化学吸附也是植物吸收砷的一种途径（Robinson et al.，2006；Azizur Rahman et al.，2008）。

水生生物对砷的富集能力与其在食物链中的营养级、摄食行为及栖息地密切相关（Watanabe et al.，2008；Culioli et al.，2009；Cott et al.，2016；Gedik et al.，2017；Liu et al.，2018）。大量研究都表明，底栖无脊椎生物中的砷含量高于鱼类（Schaeffer et al.，2006；Culioli et al.，2009；Roig et al.，2013；Fliedner et al.，2014；Juncos et al.，2016；Yang et al.，2017）。

在无脊椎动物中,砷含量依次为贻贝类>田螺类>虾类(Zhang et al., 2013)。淡水鱼的砷含量相对较低,一般小于 1 mg/kg(Ikemoto et al., 2008; de Rosemond et al., 2008; Ciardullo et al., 2010; Fu et al., 2011; Alamdar et al., 2017; Liu et al., 2018)。鱼的不同器官与组织对砷的富集能力不同,一般表现为肝脏中的砷含量最高,而鱼肉中的砷含量较低(Maher et al., 1999; Suhendrayatna et al., 2001b; Schaeffer et al., 2006; de Rosemond et al., 2008; Shah et al., 2009a; Revenga et al., 2012; Has-Schön et al., 2015; Cott et al., 2016; Dovick et al., 2016; Perera et al., 2016; Juncos et al., 2016; Yang et al., 2017)。

相比海洋生物,关于淡水水生生物中的砷形态特征的报道并不多见,相关研究更为复杂和多样。目前,关于淡水水生生物中的砷形态研究并没有一致的结论。有研究发现,淡水水生生物中的砷形态与海洋生物一致,主要为 AsB(Šlejkovec et al., 2004; Miyashita et al., 2009; Ciardullo et al., 2010; Ruttens et al., 2012; Hong et al., 2014)。但是也有研究表明,淡水水生生物中的砷形态主要为 DMA(Šlejkovec et al., 2004; Jankong et al., 2007; Miyashita et al., 2009; Cott et al., 2016; Yang et al., 2017)。大多数研究表明,淡水鱼类中的砷形态非常复杂,并非单一砷形态占绝对优势,通常由多种砷形态组成,不同物种间的砷形态差异较大(de Rosemond et al., 2008; Williams et al., 2009; Hong et al., 2014)。

1.2.3　食物链(网)中砷的富集与转化效应

1.2.3.1　砷的生物富集效应

重金属污染物在环境中进行迁移时,一旦进入食物链,就可能在生物体中富集和转化（Azizur Rahman et al.，2012）。生物累积（Bioaccumulation）是指生物体在生长发育过程中，直接通过环境和食物累积某些元素或难以分解的化合物的过程。生物富集（Bioconcentration）是指生物体或同一营养级上的多种生物种群，从所栖息的环境中蓄积化学物质，使生物体内化学物质浓度大于环境中的浓度的过程，常用生物富集系数（BF，Bioaccumulation Fator）来表征。生物转化（Biotransformation）是指污染物进入生物体后，通过肝脏或其他组织等发生化学结构改变或价态变化的过程。生物放大（Biomagnification）是指生物体内某种元素或难分解化合物的浓度随生态系统中食物链营养级的提高而逐步增大的现象；而生物稀释（Biodilution）则是指随着营养级的提高，污染物浓度在生物体内逐步降低的现象。

自20世纪50年代日本"水俣汞污染事件"报道以来，重金属沿食物链的传递受到越来越多的关注。与汞不同，而与其他重金属元素相似，砷更多地富集于食物链的底端，即生产者和初级消费者中，在高级消费者中砷富集程度趋于降低。Chen和Folt（2000）对Aberjona Watershed中的浮游动物和6种鱼类的研究表明，砷在浮游动物中含量最高，而随着营养级的增加砷含量降低。Culioli等人（2009）对Bravona河中的水生生物体中砷含量的研究发现，水生植物>苔藓植物>鱼。Cui等人（2011）发现，在黄河三角洲，植物、虾、蟹、鱼和水鸟等生物中的砷含量随着营养级的升高而降低。Vizzini等人（2013）发现，在半封闭的沿海地区，藻类、水草、鱼和鸟等不同生物中砷含量随着营养级的升高而降低。Revenga等人（2012）发现，在Moreno湖东部和西部的浮游植物、浮游动物、鱼类中，砷随着营养级的升高呈下降趋势。Foust等人（2016）对Montezuma Well食物网中的砷的研究也发现，砷含量随着营养级的增加而降低。

室内模拟实验也证实了生物体中砷含量随食物链营养级的提高而显著减少。Giddings和Eddlemon（1977）首次发现砷在不同营养级生物中的关系：在浮游植物、浮游动物、蜗牛中，砷的富集系数依次降低，分别为965、192和11。同样，大量室内实验结果表明，在藻类、浮游动物、虾、鱼中，砷含量依次下降（Maeda et al.，1992，1993；Kuroiwa et al.，1994；Suhendrayatna et al.，2001a，2002）。近年来，氮同位素是分析生态系统中生物营养级的重要手段（Fry，1988；Moriarty et al.，2009；Saigo et al.，2015）。在此基础上，很多研究也表明，水生生物中的砷含量与氮同位素含量呈明显的负相关关系（Pereira et al.，2010；Revenga et al.，2012；Vizzini et al.，2013；Juncos et al.，2016）。

但是，也有研究表明，自然环境中生物体对重金属的富集不一定和生物在食物链中所处的营养级相关（Watanabe et al.，2008；Zhang and Wang，2012；Hepp et al.，2017）。有的研究甚至相反，发现生物中的砷含量与氮同位素含量呈正相关关系（Ikemoto et al.，2008；Liu et al.，2018）。总的来说，生物对砷的富集能力的影响因素很多，主要包括生物自身（代谢机制、组织特异性、生长发育阶段、生态型、摄食、同化和排出等）和外界条件（环境浓度、暴露时间、重金属形态、环境理化性质等）两方面的因素（Ali et al.，2009；Zhang et al.，2013）。由于砷具有毒性，高砷污染可导致生态系统中食物链结构的改变（Watanabe et al.，2008；Chen et al.，2015）。生物在生长发育过程中，其摄食规律（食源和食性）会发生转变，如某些水生生物的栖息地、季节等发生变化时，其摄食对象也会发生转换，由此影响重金属的生物富集效应（Cui et al.，2011；余杨 等，2013；韦丽丽 等，2016）。

1.2.3.2 砷的代谢与转化机制

研究表明，不同类型的生物（包括原核生物、真核生物、哺乳动物和高等植物等）对砷的富集和代谢机制不同。砷的甲基化是砷同生物细胞中蛋白质结合的过程，改变酶蛋白和非酶蛋白的结构和活性，进而对生物体产生毒害作用（Shen et al.，2013；Ghosh et al.，2013）。甲基砷可在生物体内通过对无机砷的生物转化作用产生，并缓慢排出体外（Azizur Rahman and Hassler，2014）。例如，骨条藻和根管藻能够吸收 As(Ⅴ)并将其转化为 As(Ⅲ)和 DMA(Ⅴ)，而蓝隐藻则可以产生 MMA(Ⅴ)，然后释放到水体中（Azizur Rahman and Hasegawa，2012）。张兵等人（2011）对单细胞藻类（集胞藻）的研究发现，只有高浓度砷才能诱导该藻体内砷的甲基化机制起作用。对于高等生物，Braeuer 等人（2017）首次对大熊猫转化砷的行为进行研究，发现大熊猫将无机砷转化为甲基砷并将其排出体外的能力远高于其他陆地哺乳动物。

尽管已有一些对食物链中砷富集传递作用的报道，但研究尚未关注不同营养级生物体中砷的生物转化行为与富集之间的关系（Maeda et al., 1993；Chen and Folt，2000；Culioli et al.，2009；Mogren et al.，2013）。关于食物链中不同营养级的生物对砷的富集和转化的研究仅有一些室内模拟实验报道。Maeda 等人（1993）通过室内砷添加实验发现，藻类（*Nostoc sp.*）、虾（*Neocaridina denticulata*）、肉食性鲤鱼（*Cyprinus carpio*）中的甲基砷占比从低于 1%增加至 21%。Kuroiwa 等人（1994）通过室内砷添加实验也发现，藻类（*Chlorella vulgaris*）、虾（*Neocaridina denticulata*）、鳉鱼（*Oryzias latipes*）中的甲基砷占比依次从 10.9%增加至 80%。Suhendrayatna 等人（2001a）通过室内砷添加实验发现，藻类（*Chlorella vulgaris*）、浮游动物（*Daphnia magna*）、虾（*Neocaridina denticulata*）、肉食性鱼类（*Tilapia mossambica*）中甲基砷含量也依次增加。因此，是否与甲基汞相似，甲基砷由于其亲脂性会在食物链中不断富集，对生物体产生危害乃至影响食物链的结构？这个问题需要开展更多的研究才能回答。

1.2.4　稳定同位素确定营养级方法概述

相对于具有复杂结构的食物网而言，食物链常常被用来综合指示能量从食物网底端的初级生产者、有机碎屑到顶端捕食者之间的流动（Post and Takimoto，2007）。生物群落的组成成分不同，形成了各种不同的食物链。营养级关系是群落内各生物成员之间最重要的联系，是群落赖以生存的基础，也是了解生态系统能量流动的核心（张丹 等，2010）。一般来说，食物链越长，营养级的数目越多。最简单的营养结构包含两个营养级，较复杂的营养结构包含5个营养级，超过5个营养级的相对少见。通常采用所研究群落中所有物种中最高营养级物种的位置作为该群落的食物链长度（Food Chain Length，FCL）（Post，2000a；Post and Takimoto，2007）。食物链长度是生态系统最重要的特点之一，它通过影响群落结构、物种多样性、生物间营养交互作用，以及生态系统内部群落的稳定性，进而改变生态系统的主要功能（张欢 等，2013），例如，通过影响污染物生物富集（Tomy et al.，2009），从而决定生物体内污染物的浓度（Post，2002a），并且也很大程度上决定了顶级捕食者的污染物浓度（Kidd et al.，1998）。

传统的研究动物食性的方法主要是对动物消化道内的食物成分（肠含物或是胃内容物）进行分析，但这种方法主要反映生物被捕捉前短期内的摄食情况。胃内容物中通常是难以消化的食物，实际消化吸收的物质难以分辨，且采集到的很多样品中胃内容物已排空或半排空，影响研究人员对食物的鉴定（王玉玉 等，2009）。同时，在研究小型动物方面存在困难，食性分析结果往往需要校正，因此必须增大样品基数，工作量大（Kling et al.，1992；Yoshioka et al.，1994；Beaudoin et al.，1999；黎道洪、苏晓梅，2012）。近年来，稳定同位素的比值能够反映研究对象长期消化吸收的食物来源、营养位置和食物网结构，可有效用于分析生态系统中的营养流动和生物之间的营养关系（Overman and Parrish，2001；Philips et al.，2005）。$\delta^{13}C$ 和 $\delta^{15}N$ 在两个营养级间的差值常常为0‰～1‰和3‰～4‰（Peterson and Fry，1987；Vander Zanden and Rasmussen，1999，2001；Vander Zanden et al.，1997，1999；Post，2002b）。因此，在顶级捕食物种体内，$\delta^{13}C$ 和 $\delta^{15}N$ 的浓度是最高的。捕食者体内的碳同位素比值与其食物中的碳同位素比值接近，因此碳同位素比值可以用来研究物种的食物来源（Deniro and Epstein，1977）；而捕食者体内氮同位素比值与其食物中的氮同位素比值有明显不同，因此氮同位素比值可以用来研究其营养级和营养结构（Deniro and Epstein，1981；Minagawa and Wada，1984）。目前，国内外应用稳定同位素技术分析生态系统食物网结构的研究相对比较广泛（Ikemoto et al.，2008；Watanabe et al.，2008；Marín-Guirao et al.，2008；王玉玉 等，2009；Pereira et al.，2010；Cui et al.，2011；李斌，2012；Revenga et al.，2012；Zhang and Wang，2012；Wang et al.，2012；Luo et al.，2013；Vizzini et al.，2013；Gardea-Torresdey et al.，2014；郑新庆 等，2015；Saigo et al.，2015；Yadav et al.，2015；Liu et al.，2018）。

1.2.5　砷形态分析方法研究进展

目前，高效液相色谱—电感耦合等离子体质谱 HPLC-ICP-MS 由于检测限低、分离能力强、灵敏度高、分析线性范围广，已成为砷形态测试方面的主要手段（Chen et al.，2014）。但是，传统的甲醇作为提取液，提取率低，同时受检测限制，不能全面分析样品中的砷形态（Smith et al.，2008；Moriarty et al.，2009）。近年来，X 射线吸收光谱，如近边吸收光谱（XANES），对于液相和固相样品都不需要预处理，可以直接获取元素的氧化态和原子结构，目前已逐渐成为环境和生物领域中进行形态分析的重要手段（Parsons et al.，2002；Smith et al.，2005；Moriarty et al.，2009；Whaley-Martin et al.，2012）。

1.2.5.1　HPLC-ICP-MS 测试砷形态

HPLC-ICP-MS 是砷形态测试方面的主要手段。不同形态的砷化合物具有不同的稳定性，样品基质也较复杂，因此有必要选用适当的前处理技术,确保良好回收率的同时保持砷在样品中的原始形态,以提高结果的可靠性和准确性(Sadee et al.,2015)。在提取砷化物的过程中,按照提取剂的不同可分为酶提取、水提取、甲醇—水提取、氯仿—甲醇—水提取和Tris-HCl提取等(Santos et al.,2013)。由于生物样品中有机物较多,砷的形态以有机砷为主,一般采用一定比例的甲醇—水溶液提取,以1∶9、1∶1和9∶1运用最多(Maher et al.,1999;Moriarty et al.,2009;Niegel and Matysik,2010)。Ciardullo等人(2010)利用体积比为1∶1的甲醇—水溶液提取意大利Tiber河中多种鱼组织中的砷形态时,提取率可达到64%~89%。而Šlejkovec等人(2004)采用体积比为9∶1的甲醇—水溶液作为萃取剂对斯洛文尼亚的4种鱼进行砷形态分析,发现鲑鱼和鳕鱼的提取效果相对较好(53%~100%),而鲤鱼和鲶鱼的提取率相对较低(0.4%~36%)。Peng等人(2014)采用酶提取进行鸡肝脏中砷形态分析,发现在胰蛋白酶辅助提取下洛克沙砷(ROX)的回收率为46%左右,而采用胃蛋白酶则高达90%。水—乙酸溶液因其pH值与标准溶液相近,操作简单,也常常被用来作为砷形态的提取剂(吕超 等,2010)。另外,为消除样品中脂肪的影响,往往采用丙酮去除脂肪后再进行砷形态的提取(Maher et al.,1999;de Rosemond et al.,2008)。也有大量研究表明,采用纯水提取鱼类和贝类中的砷形态远远优于甲醇—水提取(Leufroy et al.,2011;Ruttens et al.,2012)。一般来说,许多砷形态都是水溶性的,在水中就能被提取出来;另外,甲醇会影响色谱分离,而导致峰形扭曲失真,因此需要用氮吹或旋转蒸发的方法加以去除(Gómezariza et al.,2000),样品中的部分砷在除去甲醇的过程中可能损失掉(Gómezariza et al.,2000;Francesconi and Kuehnelt,2004)。

  土壤和沉积物中的砷通常以无机砷形态存在,有机砷含量仅占很少一部分(Huang and Matzner,2007)。目前分析土壤和沉积物中的砷形态所采取的提取剂种类有很多,有水、甲醇—水、柠檬酸铵—盐酸、磷酸、EDTA、HCl、草酸铵、碳酸钠及碳酸氢钠等提取剂,其中最常用的是磷酸(Thomas et al.,1997;McKiernan et al.,1999;吕超 等,2010)。大米中的砷形态所采取的提取剂也多种多样，主要有水、HNO_3、HNO_3-H_2O_2、甲醇—水、酶等及其混合溶液（Narukawa et al.，2008；Welna et al.，2015）。淡水藻类和海洋藻类中的砷形态提取方法也不尽相同，有水、甲醇—水，氯仿—甲醇等提取剂（Wang et al.，2015）。因此，对于不同基质中砷形态的提取至今没有一个统一的标准，有很多因素需要考虑。Pizarro 等人（2003）用水、甲醇—水（体积比 1∶1、9∶1、1∶1～9∶1 连续提取）、磷酸进行砷形态提取，发现体积比 1∶1 的甲醇—水提取大米、鱼和鸡组织中的砷形态效率高，1 mol/L 磷酸提取土壤中的砷形态效率高且操作简单。为了结果的准确性和可靠性，应采用多种提取剂提取砷形态，逐步进行方法优化，同时可以采用标准物质对砷形态分析进行质量控制，主要有角鲨肌肉组织（DORM-2）(Schaeffer et al.，2006；Huang and Matzner，2007；de Rosemond et al.，2008）、龙虾肝胰腺（TORT-2）（Hirata et al.，2006；余晶晶 等，2009）、金枪鱼（BCR-627）（Ruttens et al.，2012）等；在标准物质缺乏的情况下，也可以通过加标回收法进行砷形态提取分析质量控制。

值得注意的是，由于实际样品提取液的 pH 值与标准溶液的 pH 值有差异，用酸做提取剂会导致砷形态的保留时间产生一定的偏移（吕超 等，2010）；HNO_3 提取样品的过程会将部分 As(Ⅲ)氧化为 As(Ⅴ)，使砷形态发生部分转变（Mccleskey et al.，2004；Kumar and Riyazuddin，2010）；甲醇—水虽较少改变样品中的砷形态，但是在样品提取之后，提取液还需要氮吹浓缩，操作烦琐，且浓缩时间较长，很难避免砷形态的损失（Huang and Ilgen，2004）；在前处理过程中加入的任何化合物或离子都可能对化学形态分析产生干扰（张磊，2007）。因此，每分析完一件样品后，用超纯水等冲洗色谱柱，可有效去除干扰物质在色谱柱上的累积。另外，鉴于大多数高分子无机盐和有机化合物在-15℃乙醇中会析出，在前处理过程中可加入一定量的纯乙醇（吕超 等，2010）。

砷形态提取方法按照使用辅助仪器的不同可分为震荡提取、超声提取、微波提取及索式提取等（Santos et al.，2013）。震荡提取和超声提取属于砷形态提取的传统方法，可以通过增加提取剂与样品的接触面积，从而快速溶出内部物质（B'Hymer and Caruso，2004）。但近年来，微波提取可以通过设定功率、时间、温度、压强等达到很好的提取效果，被广泛使用（Hirata et al.，2006）。Narukawa 等人（2008）用 HPLC-ICP-MS 检测大米面粉中的砷形态，通过对比震荡提取、超声提取和微波提取等方式发现，采用微波提取效果最好，提取率为 97%~106%，过程为：从室温升至 80℃（5 min），保持 30 min，再冷却至室温（10 min）（Sadee et al.，2015）。Ruttens 等人（2012）应用同样的方法检测 350 多种食品样品，发现大多数海洋鱼类中砷形态提取率为 78%~110%，但是个别淡水鱼中砷形态的提取率很低，仅为 37%~54%。加速溶剂萃取方法（ASE）是一种相对较新的提取技术，通过调节温度、提升压强，保证溶剂低于其沸点，从而实现安全和快速提取（Sadee et al.，2015）。但研究发现，利用 ASE 提取方法，样品必须分散于惰性介质中，否则砷形态可能会大量转化，同时会增加其他有机物而导致色泽变深，对砷形态分析产生干扰。

阴离子交换柱主要选用 Hamilton PRP-X100，文献报道的流动相大多为磷酸盐和碳酸盐两种，包括$(NH_4)_2CO_3$（Ciardullo et al.，2010）、NH_4HCO_3（Schaeffer et al.，2006），以及 $NH_4H_2PO_4$（Maher et al.，1999；Zheng et al.，2003；Schaeffer et al.，2006；Jankong et al.，2007；Price et al.，2016）。还有报道采取梯度洗脱，如$(NH_4)_2CO_3$、$(NH_4)_3PO_4$ 和 NH_4NO_3（Hong et al.，2014），以及$(NH_4)_2CO_3$、H_2O（Ruttens et al.，2012）。不同流动相的 pH 值变化较大，因此，在实际操作过程中，应根据实验条件进行调试和优化。以往研究 ICP-MS 主要以 ^{75}As 作为检测信号，但是由于样品中的某些元素（如 Ca、S、Cl）和仪器所用的高纯载气（氩气）易形成 $^{40}Ar^{35}Cl^+$、$^{40}Ar^{34}SH^+$、$^{40}Ca^{35}Cl^+$、$^{37}Cl_2H^+$，造成干扰，还会受到 Nd^{2+}、Eu^{2+}、Sm^{2+} 等多种离子的干扰（Guo et al.，2011；An et al.，2017）。但是在 DRC 模式下，当 O_2 的流速超过 0.5 mL/min 时，As 迅速转化为 AsO，且转化率在 90% 以上，稳定性极高（Guo et al.，2011）。同时干扰元素 $^{91}Zr^+$ 在环境中的含量极低，

且极易被氧化，Zr：ZrO 远小于 0.04，可忽略不计（Guo et al.，2011）。因此联机后选择 ^{91}(AsO$^+$)作为检测信号的方法更具有可靠性和准确性。

1.2.5.2　XAFS 测试砷形态

砷形态分析的传统方法主要为离子色谱、气相色谱、高效液相色谱、液相色谱和离子交换柱等分离手段与元素特征检测器（火焰或石墨炉原子吸收、ICP 发射光谱或质谱、氢化物发生原子荧光等）联用，但是这些色谱方法需要对目标样品进行复杂的提取和分离等预处理，而这些处理过程可能导致砷形态发生转变和损失，影响分析结果的可信度（Paktunc et al.，2003；Hokura et al.，2006）。伴随着同步辐射技术的发展，X 射线吸收精细结构谱（X-ray Adsorption Fine Spectroscopy，XAFS）逐渐发展为一种具有较高应用价值的形态分析方法。其原理是通过透射方法或荧光方法获得样品中目标元素在某吸收边附近和高于吸收边的吸收系数和入射 X 射线能量的关系曲线。XAFS 方法不需要预分离或化学预处理过程，不会破坏其化学形态，可以直接对复杂基质样品进行无损分析，获取整个样品中目标元素的氧化态、近边原子和配位数等化学形态信息，可以实现形态分析和结构鉴定同步进行（Polette et al.，2000）。因此，XAFS 方法已成为环境和生物领域进行形态分析的重要手段（George et al.，2009；Castillo-Michel et al.，2011；Niazi et al.，2011；Zeng et al.，2015）。一条吸收谱线通常被分为两个部分：X 射线吸收近边结构谱（X-ray Adsorption Near-Edge structure，XANES）和 X 射线吸收扩展边精细结构谱（Extended X-ray Adsorption Fine Structure，EXAFS）。前者能量范围为$-20 \sim 50 \mathrm{eV}$，后者为 $50 \sim 1000 \mathrm{eV}$。XANES 能够提供吸收原子的化学形态及几何结构信息，具有"指纹效应"，而 EXAFS 则能提供吸收原子的近邻配位结构信息（Gardea-Torresdey et al.，2005）。

目前，XAFS 主要在土壤和底泥的砷形态分析中运用。李士杏等人（2011）采用 XANES 分析发现，在小于 5 μm 的红壤中，砷形态主要与针铁矿和无定形氧化铁相结合。陈学萍等人（2008）采用 XAFS 研究发现，用硝酸钾和硫酸钾处理后的水稻，其根系表面铁膜中的砷形态主要为三价砷。Tang 等人（2016）采用 XANES 对石门雄黄矿区的土壤进行分析，得出其砷形态主要为砷酸盐。值得一提的是，X 射线光谱检测法只能分析砷含量高的样品，例如，在砷超富集植物如蜈蚣草中的运用颇多。万小铭等人（2015）通过研究发现，蜈蚣草根部和羽片中以 As(V)、As(III)为主，还存在少量的 AsS。Lombi 等人（2002）发现蜈蚣草叶片中的砷以 As(III)为主。黄泽春等人（2003）采用 EXAFS 分析大叶井口边草的砷形态，结果表明其砷主要与氧配位，根部也存在与谷胱甘肽（GSH）结合的砷。Caumette 等人（2011）用 XANES 分析发现，浮游生物中的砷形态主要为 As(V)，其次为硫配位砷及砷糖。XAFS 不仅可以应用在植物中，也可分析动物中的砷形态。Hong 等人（2014）采用 μ-XANES 对 HPLC-ICP-MS 测定砷形态时的未知峰进行进一步确定。Foust 等人（2016）则采用 XANES 对美国 Montezuma Well 生态系统中的水蛭进行分析，得出其体内的砷主要为硫配位砷。上述研究表明，XAFS 技术成功地应用于不同介质中的砷形态研究，是目前唯一一种用于砷形态分析的原位分析方法，为揭示砷在环境中的行为机制和毒性提供了技术支撑。

1.3 砷矿区食物链中砷的生物富集与转化研究思路

1.3.1 砷矿区食物链中砷的生物富集与转化研究目的

目前,关于陆生和水生食物链生物体中的砷形态转化与生物代谢及食物链传递之间的关系还有很多不明确之处。石门雄黄矿砷污染时间长、范围大、程度深,是研究砷在食物链中生物富集与转化的理想地区。土壤中高含量的砷能通过食物链富集最终对人类健康造成威胁。环境介质中的砷形态主要为无机砷,但是生物体中的砷形态复杂多样,一直是国内外研究的热点和难点。矿区生长的生物接受长期高浓度的砷暴露,其体内砷含量会显著升高,因此有利于提高砷形态的检测效率。同时该矿区属于单一砷污染,能保证结果的准确性、典型性。本研究通过对石门雄黄矿区不同砷污染区多介质中砷含量及砷形态的测定,结合动物系统分类和氮同位素划分营养级,查明砷矿区砷的污染状况及分布特征,探究长期砷暴露下矿区食物链中砷的生物富集与转化行为;以生物体中无机砷的还原、甲基化等生物转化作用过程为重点，探索高砷矿区典型食物链上砷的生物转化过程与营养级的关系。通过本研究，有望揭示砷在陆生和水生食物链中的富集和转化过程，为阐明有机砷化物对生物和食物链的作用和砷的生态风险评估提供更有价值的数据和信息。

1.3.2 砷矿区食物链中砷的生物富集与转化研究内容

（1）砷矿区环境介质中砷的污染范围、程度及时空分布规律。

采集砷矿区的天然水体、表层沉积物及表层土壤样品，分别测定其中砷含量水平，了解矿区环境中砷的污染现状、空间分布及迁移方式，判定环境中砷的来源，了解砷向环境中释放的动力学及控制机制。

（2）砷矿区不同食物链中砷的生物富集行为。

采集不同食物链各营养级生物样品，将动物系统分类方法和氮、碳同位素方法相结合，精确确定各生物的食物链营养级，分析砷在食物链各营养级对应的生物和环境中的含量，揭示环境和自身生长因素对砷富集的影响作用，确定砷矿区中砷是否有明显的生物累积及食物链放大效应。

（3）砷矿区不同食物链中砷的生物转化行为。

通过比较不同营养级的生物中的砷形态特征，研究陆生环境和水生环境中的砷循环过程，阐明砷的生物地球化学行为；以生物体中无机砷的还原、甲基化等生物转化作用过程为重点，探索高砷矿区典型食物链砷的生物转化过程与营养级的关系；对比陆生生态系统和水生生态系统中砷的迁移转化，对砷的生物富集和转化进行全面的认识。

参 考 文 献

仇荣亮, 仇浩, 雷梅, 等. 矿山及周边地区多金属污染土壤修复研究进展[J]. 农业环境科学学报, 2009, 28:1085-1091.

陈国梁. 沉水植物对砷的富集特征及机理研究[D]. 浙江大学, 博士论文, 2014.

陈怀满. 环境土壤学[M]. 北京, 科学出版社, 2005.

陈同斌, 韦朝阳, 黄泽春, 等. 砷超富集植物蜈蚣草及其对砷的富集特征[J]. 科学通报, 2002, 47:207-210.

陈学萍, 朱永官, 洪米娜, 等. 不同施肥处理对水稻根表铁和砷形态的影响[J]. 环境化学, 2008, 27:231-234.

黄芳. 石门砷殇[J]. 安全与健康, 2014, 400:34-35.

黄益宗, 郝晓伟, 雷鸣, 等. 重金属污染土壤修复技术及其修复实践[J]. 农业环境科学学报, 2013, 32:409-417.

黄泽春, 陈同斌, 雷梅, 等. 砷超富集植物中砷形态及其转化的 EXAFS 研究[J]. 中国科学（C 辑）, 2003, 33:488-494.

和莉莉, 李冬梅, 吴钢. 我国城市土壤重金属污染研究现状和展望[J]. 土壤通报, 2008, 39:1210-1216.

李斌. 三峡库区小江鱼类食物网结构、营养级关系的 C、N 稳定性同位素研究[D]. 西南大学, 博士论文, 2012.

黎道洪, 苏晓梅. 应用稳定同位素研究广西东方洞食物网结构和营养级关系[J]. 生态学报, 2012, 32:3497-3504.

李慧, 张立实. 砷的毒性与生物学功能[J]. 现代预防医学, 2000, 27:39-40.

李莲芳, 曾希柏, 白玲玉, 等. 石门雄黄矿周边地区土壤砷分布及农产品健康风险评估[J]. 应用生态学报, 2010, 21:2946-2951.

李士杏, 骆永明, 章海波, 等. 红壤不同粒级组分中砷的形态——基于连续分级提取和 XANES 研究[J]. 环境科学学报, 2011, 31:2733-2739.

吕超, 刘丽萍, 董慧茹, 等. 高效液相色谱—电感耦合等离子体质谱联用技术测定水产类膳食中 5 种砷形态的方法研究[J]. 分析测试学报, 2010, 29:465-468.

蒲含勇, 张应红. 论我国矿产资源的综合利用[J]. 矿产综合利用, 2001, 4:19-22.

万小铭, 刘颖茹, 雷梅, 等. 不同种群蜈蚣草中砷形态的 X 射线吸收光谱研究[J]. 光谱学与光谱分析, 2015, 35:2329-2332.

王萍, 王世亮, 刘少卿, 等. 砷的发生、形态、污染源及地球化学循环[J]. 环境科学与技术, 2010, 33:96-103.

王玉玉, 于秀波, 张亮, 等. 应用碳、氮稳定同位素研究鄱阳湖枯水末期水生食物网结构[J]. 生态学报, 2009, 29:1181-1188.

韦朝阳, 陈同斌, 黄泽春, 等. 大叶井口边草——一种新发现的富集砷的植物[J]. 生态学报, 2002, 22:777-778.

韦丽丽, 周琼, 谢从新, 等. 三峡库区重金属的生物富集、生物放大及其生物因子的影响[J]. 环境科学, 2016, 37:325-334.

肖细元, 陈同斌, 廖晓勇, 等. 中国主要含砷矿产资源的区域分布与砷污染问题[J]. 地理研究, 2008, 27:201-212.

徐红宁, 许嘉琳. 我国砷异常区的成因及分布[J]. 土壤, 1996, 02:80-84.

杨芬, 朱晓东, 韦朝阳. 陆地水环境中砷的迁移转化[J]. 生态学杂志, 2015, 34:1448-1455.

余杨, 王雨春, 周怀东, 等. 三峡水库蓄水初期鱼体汞含量及其水生食物链累积特征[J]. 生态学报, 2013, 33:4059-4067.

余晶晶, 曹煊, 崔维刚, 等. 高效液相色谱—电感耦合等离子体质谱测定浒苔中砷及砷化学形态[J]. 食品科学, 2009, 30:223-227.

张兵, 王利红, 徐玉新, 等. 集胞藻（Synechocystis sp. PCC6803）对砷吸收转化特性的初步研究[J]. 生态毒理学报, 2011, 6:629-633.

张丹, 闵庆文, 成升魁, 等. 应用碳、氮稳定同位素研究稻田多个物种共存的食物网结构和营养级关系 [J]. 生态学报, 2010, 30 (24): 6734-6740.

张欢, 何亮, 张培育, 等. 食物链长度理论研究进展[J]. 生态学报, 2013, 33:7630-7643.

张磊, 吴永宁, 赵云峰. 不同形态砷化合物稳定性研究和砷形态分析中样品前处理技术[J]. 国外医学（卫生学分册）, 2007, 34:238-244.

张楠, 韦朝阳, 杨林生. 淡水湖泊系统中砷的赋存与转化行为研究进展[J]. 生态学报, 2013, 33:337-347.

郑新庆, 王倩, 黄凌风, 等. 基于碳、氮稳定同位素的厦门筼筜湖两种优势端足类食性分析[J]. 生态学报, 2015, 35:7589-7597.

ALAMDAR A, EQANI SAMAS, HANIF N, et al. Human exposure to trace metals and arsenic via consumption of fish from river Chenab, Pakistan and associated health risks[J]. Chemosphere, 2017, 168:1004-1012.

ALI W, ISAYENKOV SV, ZHAO FJ, et al. Arsenite transport in plants[J]. Cell Mol Life Sci, 2009, 66:2329-2339.

ALVAREZ A, ORDÓÑEZ A, LOREDO J. Geochemical assessment of an arsenic mine adjacent to a water reservoir (León, Spain)[J]. Environ Geol, 2006, 50:873-884.

AN J, LEE H, NAM K, et al. Effect of methanol addition on generation of isobaric polyatomic ions in the analysis of arsenic with ICP-MS[J]. Microchem

J, 2017, 131:170-173.

APOSHIAN HV, GURZAU ES, LE XC, et al. Occurrence of monomethylarsonous acid in urine of humans exposed to inorganic arsenic[J]. Chem Res Toxicol, 2000, 13:693-697.

AZIZUR RAHMAN M, HASEGAWA H, UEDA K, et al. Arsenic accmulation in duckweed (*Spirodela polyrhiza* L.): a good option for phytoremediation[J]. Hazard Mater, 2008, 160:356-361.

AZIZUR RAHMAN M, HASEGAWA H, LIM RP. Bioaccumulation, biotransformation and trophic transfer of arsenic in the aquatic food chain[J]. Environ Res, 2012, 116:118-135.

AZIZUR RAHMAN M, HASEGAWA H. Arsenic in freshwater systems: Influence of eutrophication on occurrence, distribution, speciation, and bioaccumulation[J]. Appl Geochem, 2012, 27:304-314.

AZIZUR RAHMAN M, HASSLER C. Is arsenic biotransformation a detoxification mechanism for microorganisms[J]. Aquat Toxicol, 2014, 146:212-219.

BARRINGER JL, SZABO Z, WILSON TP, et al. Distribution and seasonal dynamics of arsenic in a shallow lake in northwestern New Jersey, USA[J]. Environ Geochem Health, 2011, 33:1-22.

BEAUDOIN CP, TONN WM, PREPAS EE, et al. Individuals specialization and trophic adaptability of northern pike (*Esox lucius*):An isotope and dietary analysis[J]. Oecologia, 1999, 120:386-396.

BERGQVIST C, GREGER M. Arsenic accumulation and speciation in plants from different habitats[J]. Appl Geochem, 2012, 27:615-622.

B'HYMER C, CARUSO J A. Arsenic and its speciation analysis using high-performance liquid chromatography and inductively coupled plasma mass spectrometry[J]. Chromatogr A, 2004, 1045:1-13.

BRAEUER S, DUNGL E, HOFFMANN W, et al. Unusual arsenic metabolism in Giant Pandas[J]. Chemosphere, 2017, 189:418-425.

BUNDSCHUH J, NATH B, BHATTACHARYA P, et al. Arsenic in the human food chain: The Latin American perspective[J]. Sci Total Environ, 2012, 429:92-106.

BUTTON M, JENKIN GRT, HARRINGTON CF, et al. Arsenic biotransformation in earthworms from contaminated soils[J]. Environ Monitor, 2009, 11:1484-1191.

BUTTON M, MORIARTY MM, WATTS M, et al. Arsenic speciation in field-collected and laboratory-exposed earthworms *Lumbricus terrestris*[J]. Chemosphere, 2011, 85:1277-1283.

BUTTON M, KOCH I, REIMER KJ. Arsenic resistance and cycling in earthworms residing at a former gold mine in Canada[J]. Environ Pollut, 2012, 169:74-80.

CAO X, MA LQ, TU C. Antioxidative responses to arsenic in the arsenic-hyperaccumulator Chinese brake fern (*Pteris vittata L.*)[J]. Environ Pollut, 2004, 128:317-325.

CASTILLO-MICHEL H, HERNANDEZ-VIEZCAS J, DOKKEN KM, et al. Localization and speciation of arsenic in soil and desert plant *Parkinsonia florida* using μXRF and μXANES[J]. Environ Sci Technol, 2011, 45:7848-7854.

CAUMETTE G, KOCH I, ESTRADA E, et al. Arsenic speciation in plankton organisms from contaminated lakes: transformations at the base of the freshwater food chain[J]. Environ Sci Technol, 2011, 45:9917-9923.

CAUMETTE G, KOCH I, MORIARTY M, et al. Arsenic distribution and speciation in *Daphnia pulex*[J]. Sci Total Environ, 2012, 432:243-250.

CHEN CY, FOLT CL. Bioaccumulation and diminution of arsenic and lead in a freshwater food web[J]. Environ Sci Technol, 2000, 34:3878-3884.

CHEN ZL, AKTER KF, RAHMAN MM, et al. The separation of arsenic species in soils and plant tissues by anion-exchange chromatography with inductively coupled mass spectrometry using various mobile phases[J]. Microchem J, 2008, 89:20-28.

CHEN ML, MA LY, CHEN XW. New procedures for arsenic speciation: A review[J]. Talanta, 2014, 125:78-86.

CHEN G, SHI H, TAO J, et al. Industrial arsenic contamination causes catastrophic changes in freshwater ecosystems[J]. Sci Rep, 2015, 5:17419-17425.

CIARDULLO S, AURELI F, RAGGI A, et al. Arsenic speciation in freshwater fish: Focus on extraction and mass balance[J]. Talanta, 2010, 81:213-221.

COTT PA, ZAJDLIK BA, PALMER MJ, et al. Arsenic and mercury in lake whitefish and burbot near the abandoned Giant Mine on Great Slave Lake[J]. Great Lakes Res, 2016, 42:223-232.

CUI B, ZHANG Q, ZHANG K, et al. Analyzing trophic transfer of heavy metals for food webs in the newly-formed wetlands of the Yellow River Delta, China[J]. Environ Pollut, 2011, 159:1297-1306.

CULIOLI JL, FOUQUOIRE A, CALENDINI S, et al. Trophic transfer of arsenic and antimony in a freshwater ecosystem: A field study[J]. Aquat Toxicol, 2009, 94:286-293.

DADA EO, NJOKU KL, OSUNTOKI AA, et al. Evaluation of the response of a wetland, tropical earthworm to heavy metal contaminated soil[J]. Environ Monit Anal, 2013, 1:47-52.

DE ROSEMOND S, XIE Q, LIBER K. Arsenic concentration and speciation in five freshwater fish species from Back Bay near Yellowknife, NT, Canada[J]. Environ Monit Assess, 2008, 147:199-210.

DENIRO MJ, EPSTEIN S. Mechanism of carbon isotope fractionation associated with lipid synthesis science[J]. Science, 1977, 197:261-263.

DENIRO MJ, EPSTEIN S. Influence of diet on the distribution of nitrogen isotopes in animals[J]. Geochim Cosmochim Ac, 1981, 45:341-351.

DOVICK MA, KULP TR, ARKLE RS, et al. Bioaccumulation trends of arsenic and antimony in a freshwater ecosystem affected by mine drainage[J]. Environ Chem, 2016, 13:149-159.

FLIEDNER A, RÜDEL H, KNOPF B, et al. Spatial and temporal trends of metals and arsenic in German freshwater compartments[J]. Environ Sci Pollut Res, 2014, 21:5521-5536.

FOUST JR RD, BAUER AM, COSTANZA-ROBINSON M, et al. Arsenic transfer and biotransformation in a fully characterized freshwater food web[J]. Coordin Chem Rev, 2016, 306:558-565.

FRANCESCONI KA, KUEHNELT D. Determination of arsenic species: A critical review of methods and applications, 2000-2003[J]. Analyst, 2004, 129:373-395.

FRY B. Food web structure on Georges Bank from stable C, N, and S isotopic compositions[J]. Limnol. Oceanogr, 1988, 33:1182-1190.

FU Z, WU F, AMARASIRIWARDENA D, et al. Antimony, arsenic and mercury in the aquatic environment and fish in a large antimony mining area in Hunan, China[J]. Sci Total Environ, 2010, 408:3403-3410.

FU Z, WU F, MO C, et al. Bioaccumulation of antimony, arsenic, and mercury in the vicinities of a large antimony mine, China[J]. Microchem J, 2011, 97:12-19.

GARCÍA-GÓMEZ C, ESTEBAN E, SÁNCHEZ-PARDO B, et al. Assessing the ecotoxicological effects of long-term contaminated mine soils on plants and earthworms: relevance of soil (total and available) and body concentrations[J]. Ecotoxicology, 2014, 23:1195-1209.

GARDEA-TORRESDEY JL, PERALTA-VIDEA JR, ROSA GDL, et al. Phytoremediation of heavy metals and study of the metal coordination by X-ray absorption spectroscopy[J]. Coordin Chem Rev, 2005, 249:1797-1810.

GARDEA-TORRESDEY JL, RICO CY, WHITE JC. Trophic transfer, transformation, and impact of engineered nanomaterials in terrestrial environments[J]. Environ Sci Technol, 2014, 48:2526-2540.

GEDIK K, KONGCHUM M, DELAUNE RD, et al. Distribution of arsenic and other metals in crayfish tissues (*Procambarus clarkii*) under different production practices[J]. Sci Total Environ, 2017, 574:322-331.

GEORGE GN, PRINCE RC, SINGH SP, et al. Arsenic K-edge X-ray absorption spectroscopy of arsenic in seafood[J]. Mol Nutr Food Res, 2009, 53:552-557.

GHOSH A, MAJUMDER S, AWAL MA, et al. Arsenic exposure to dairy cows in Bangladesh[J]. Arch Environ Con Tox, 2013, 64:151-159.

GIDDINGS JM, EDDLEMON GK. The effects of microcosm size and substrate type on aquatic microcosm behavior and arsenic transport[J]. Arch

Environ Con Tox, 1977, 6:491-505.

GOERING PL, APOSHIAN HV, MASS MJ, et al. The enigma of arsenic carcinogenesis: role of metabolism[J]. Toxicol Sci, 1999, 49:5-14.

GÓMEZARIZA JL, SÁNCHEZRODAS D, GIRÁLDEZ I, et al. Comparison of biota sample pretreatments for arsenic speciation with coupled HPLC-HG-ICP-MS[J]. Analyst, 2000, 125:401-407.

GRACE EJ, MACFARLANE GR. Assessment of the bioaccumulation of metals to chicken eggs from residential backyards[J]. Sci Total Environ, 2016, 563-564:256-260.

GREGUS Z, GYURASICS A, CSANAKY I. Biliary and urinary excretion of inorganic arsenic: Monomethylarsonous acid as a major biliary metabolite in rats[J]. Toxicol Sci, 2000, 56:18-25.

GUO W, HU S, LI X, et al. Use of ion－molecule reactions and methanol addition to improve arsenic determination in high chlorine food samples by DRC-ICP-MS[J]. Talanta, 2011, 84:887-894.

HAS-SCHÖN E, BOGUT I, VUKOVIĆ B, et al. Distribution and age-related bioaccumulation of lead (Pb), mercury (Hg), cadmium (Cd), and arsenic (As) in tissues of common carp (*Cyprinus carpio*) and European catfish (*Sylurus glanis*) from the Bu š ko Blato reservoir (Bosnia and Herzegovina)[J]. Chemosphere, 2015, 135, 289-296.

HEPP LU, PRATAS JAM, GRAÇA M. Arsenic in stream waters is bioaccumulated but neither biomagnified through food webs nor biodispersed to land[J]. Ecotox Environ Safe, 2017, 139:132-138.

HIRATA S, TOSHIMITSU H, AIHARA M. Determination of arsenic species in marine samples by HPLC-ICP-MS[J]. Analytical Sciences: *the*

International Journal of the Japan Society for Analytical Chemistry Japan Society for Analytical Chemistry, 2006, 22:39-43.

HOKURA A, OMUMA R, TERADA Y, et al. Arsenic distribution and speciation in an arsenic hyperaccumulator fern by X-ray spectrometry utilizing a synchrotron radiation source[J]. Anal Atom Spectrom, 2006, 21:321-328.

HONG S, KHIM JS, PARK J, et al. Species- and tissue-specific bioaccumulation of arsenicals in various aquatic organisms from a highly industrialized area in the Pohang City, Korea[J]. Environ Pollut, 2014, 192: 27-35.

HUANG JH, ILGEN G. Blank values, adsorption, pre-concentration, and sample preservation for arsenic speciation of environmental water samples[J]. Analytica Chimica Acta, 2004, 512:1-10.

HUANG JH, MATZNER E. Mobile arsenic species in unpolluted and polluted soils[J]. Sci Total Environ, 2007, 377:308-318.

HUANG ZC, CHEN TB, LEI M, et al. Difference of toxicity and accumulation of methylated and inorganic arsenic in arsenic-hyperaccumulating and -hypertolerant plants[J]. Environ Sci Technol, 2008, 42:5106-5111.

HUGHES MF. Arsenic toxicity and potential mechanisms of action[J]. Toxicol Lett, 2002, 133:1-16.

HUGHES MF, BECK BD, CHEN Y, et al. Arsenic exposure and toxicology: A historical perspective[J]. Toxicol Sci, 2011, 123:305-332.

IKEMOTO T, TU NPC, OKUDA N, et al. Biomagnification of trace elements in the aquatic food web in the Mekong Delta, South Vietnam using stable carbon and nitrogen isotope analysis[J]. Arch Environ Contam Toxicol, 2008, 54:504-515.

JANKONG P, CHALHOUB C, KIENZL N, et al. Arsenic accumulation and speciation in freshwater fish living in arsenic-contaminated waters[J]. Environ Chem, 2007, 4:11-17.

JEDYNAK L, KOWALSKA J, HARASIMOWICZ J, et al. Speciation analysis of arsenic in terrestrial plants from arsenic contaminated area[J]. Sci Total Environ, 2009, 407:945-952.

JEONG S, MOON HS, NAM K. Enhanced uptake and translocation of arsenic in Cretan brake fern (*Pteris cretica L.*) through siderophorearsenic complex formation with an aid of rhizospheric bacterial activity[J]. Hazard Mater, 2014, 280:536-543.

JUNCOS R, ARCAGNI M, RIZZO A, et al. Natural origin arsenic in aquatic organisms from a deep oligotrophic lake under the influence of volcanic eruptions[J]. Chemosphere, 2016, 144:2277-2289.

KIDD KA, SCHINDLER DW, HESSLEIN RH, et al. Effects of trophic position and lipid on organochlorine concentration in fishes from sub-Arctic lakes in Yukon Terriotory[J]. Can J Fish Aquat Sci, 1998, 55:869-881.

KIM YT, YOON HO, YOON C, et al. Arsenic species in ecosystems affected by arsenic-rich spring water near an abandoned mine in Korea[J]. Environ Pollut, 2009, 157:3495-3501.

KLING GW, FRY B, O'BRIEN WJ. Stable isotopes and planktonic trophic structure in arctic lakes[J]. Ecology, 1992, 73:561-566.

KOCH I, MACE JV, REIMER KJ. Arsenic speciation in terrestrial birds from Yellowknife, Northwest Territories, Canada: the unexpected finding of arsenobetaine[J]. Environ Toxicol Chem, 2005, 24:1468-1474.

KUBOTA R, KUNITO T, TANABE S. Occurrence of severak arsenic

compounds in the liver of birds, cetaceans, pinnipeds, and sea turtles[J]. Environ Toxicol Chem, 2003, 22:1200-1207.

KUMAR AR, RIYAZUDDIN P. Preservation of inorganic arsenic species in environmental water samples for reliable speciation analysis[J]. Trac Trend Anal Chem, 2010, 29:1212-1223.

KUROIWA T, OHKI A, NAKA K, et al. Biomethylation and biotransformation of arsenic in a freshwater food chain: Green alga (*Chlorella vulgaris*)→shrimp (*Neocaridina denticulata*)→killifish (*Oryzias iatipes*)[J]. Appl Organometal Chem, 1994, 8:325-333.

LANGDON CJ, PIEARCE TG, FELDMANN J, et al. Arsenic speciation in the earthworms *Lumbricus rubellus* and *Dendrodrilus rubidus*[J]. Environ Toxicol Chem, 2003a, 22:1302-1308.

LANGDON CJ, PIEARCE TG, MEHARG AA, et al. Interactions between earthworms and arsenic in the soil environment: A review[J]. Environ Pollut, 2003b,124:361-373.

LARIOS R, FERNÁNDEZ-MARTÍNEZ R, LEHECHO I, et al. A methodological approach to evaluate arsenic speciation and bioaccumulation in different plant species from two highly polluted mining areas[J]. Sci Total Environ, 2012, 414:600-607.

LEUFROY A, NOËL L, DUFAILLY V, et al. Determination of seven arsenic species in seafood by ion-exchange chromatography coupled to inductively coupled plasma mass spectrometry following microwave assisted extraction: Method validation and occurrence data[J]. Talanta, 2011, 83:770-779.

LEVY JL, STAUBER JL, ADAMS MS, et al. Toxicity, biotransformation, and mode of action of arsenic in two freshwater microalgae (*Chlorella* sp. and *Monoraphidium arciiatum*)[J]. Environ Toxicol Chem, 2005, 24:2630-2639.

LEYBOURNE M, JOHANNESSON KH, ASFAW A. Measuring arsenic speciation in environmental media: sampling, preservation, and analysis[J]. Rev Mineral Geochem, 2014,79:371-390.

LIU T, GUO H, XIU W, et al. Biomarkers of arsenic exposure in arsenic-affected areas of the Hetao Basin, Inner Mongolia[J]. Sci Total Environ, 2017, 609:524-534.

LIU Y, LIU G, YUAN Z, et al. Heavy metals (As, Hg and V) and stable isotope ratios ($\delta^{13}C$ and $\delta^{15}N$) in fish from Yellow River Estuary, China[J]. Sci Total Environ, 2018, 613-614:462-471.

LOMAX C, LIU WJ, WU L, et al. Methylated arsenic species in plants originate from soil microorganisms[J]. New Phytol, 2012, 193:665-672.

LOMBI E, ZHAO FJ, FUHRMANN M, et al. Arsenic distribution and speciation in the fronds of the hyperaccumulator *Pteris vittata*[J]. New Phytol, 2002, 156:195-203.

LUO XJ, ZENG YH, CHEN HS, et al. Application of compound-specific stable carbon isotope analysis for the biotransformation and trophic dynamics of PBDs in a feeding study with fish[J]. Environ Pollut, 2013, 176:36-41.

MA LQ, KOMAR KM, TU C, et al. A fern that hyperaccumulates arsenic[J]. Nature, 2001, 409:579-579.

MADIGAN BA, TUROCZY N, STAGNITTI F. Speciation of arsenic in a large endoheric lake[J]. Bull Environ Contam Toxico, 2005, 75:1107-1114.

MAEDA S, OHKI A, KUSADOME K, et al. Bioaccumulation of arsenic and its fate in a freshwater food chain[J]. Appl Organometal Chem, 1992,

23:213-219.

MAEDA S, MAWATARI K, OHKI A, et al. Arsenic metabolism in a freshwater food chain: Blue－green alga (*Nostoc* sp.)→shrimp (*Neocaridina denticulata*)→carp (*Cyprinus carpio*)[J]. Appl Organometal Chem, 1993, 7:467-476.

MAHER W, GOESSLER W, KIRBY J, et al. Arsenic concentrations and speciation in the tissues and blood of sea mullet (Mugil cephalus) from Lake Macquarie NEW, Australia[J]. Mar Chem, 1999, 68:169-182.

MANDAL BK, SUZUKI KT. Arsenic round the world: a review[J]. Talanta, 2002, 58:201-235.

MARÍN-GUIRAO L, LLORET J, MARIN A. Carbon and nitrogen stable isotopes and metal concentration in food webs from a mining-impacted coastal lagoon[J]. Sci Total Environ, 2008, 393:118-130.

MATSCHULLAT J. Arsenic in the geosphere-a review[J]. Sci Total Environ, 2000, 249:297-312.

MCCLESKEY RB, NORDSTROM DK, MAEST AS. Preservation of water samples for arsenic(III/V) determinations: An evaluation of the literature and new analytical results[J]. Appl Geochem, 2004, 19:995-1009.

MCKIERNAN JW, CREED JT, BROCKHOFF CA, et al. A comparison of automated and traditional methods for the extraction of arsenicals from fish[J]. Anal Ato Spectrom, 1999, 14:607-613.

MEHARG AA, HARTLEY-WHITAKER J. Arsenic uptake and metabolism in arsenic resistant and nonresistant plant species[J]. New Phytol, 2002, 154: 29-43.

MINAGAWA M, WADA E. Stepwise enrichment of $\delta^{15}N$ along food chains: Further evidence and the relation between $\delta^{15}N$ and animal age[J]. Geochim Cosmochim Ac, 1984, 48:1135-1140.

MIR KA, RUTTER A, KOCH I, et al. Extraction and speciation of arsenic in plants grown on arsenic contaminated soils[J]. Talanta, 2007, 72:1507-1518.

MIYASHITA S, SHIMOYA M, KAMIDATE Y, et al. Rapid determination of arsenic species in freshwater organisms from the arsenic-rich Hayakawa River in Japan using HPLC-ICP-MS[J]. Chemosphere, 2009, 75:1065-1073

MOGREN CL, WALTON WE, PARKER DR, et al. Trophic transfer of arsenic from an aquatic insect to terrestrial insect predators[J]. Plos One, 2013, 8:1-6.

MORIARTY MM, KOCH I, GORDON RA, et al. Arsenic speciation of terrestrial invertebrates[J]. Environ Sci Technol, 2009, 43:4818-4823.

NARUKAWA T, INAGAKI K, KUROIWA T, et al. The extraction and speciation of arsenic in rice flour by HPLC-ICP-MS[J]. Talanta, 2008, 77:427-432.

NG JC, WANG J, SHRAIM A. A global health problem caused by arsenic from natural sources[J]. Chemosphere, 2003, 52:1353-1359.

NIAZI NK, SINGH B, SHAH P. Arsenic speciation and phytoavailability in contaminated soils using a sequential extraction procedure and XANES spectroscopy[J]. Environ Sci Technol, 2011, 45:7135-7142.

NIEGEL C, MATYSIK FM. Analytical methods for the determination of arsenosugars-a review of recent trends and developments[J]. Analytical Chimica Acta, 2010, 657:83-99.

NIKOLAIDIS NP, DOBBS GM, CHEN J, et al. Arsenic mobility in contaminated lake sediments[J]. Environ Pollut, 2004, 129:479-487.

NRIAGU JO, PACYNAT JM. Quantitative assessment of worldwide contamination of air, water and soils by trace metals[J]. Nature, 1988, 333:134-139.

OTONES V, ÁLVAREZ-AYUSO E, GARCÍA-SÁNCHEZ A, et al. Arsenic distribution in soils and plants of an arsenic impacted former mining area[J]. Environ Pollut, 2011, 159:2637-2647.

OVERMAN NC, PARRISH DL. Stable isotope composition of walleye: ^{15}N accumulation with age and area-specific differences in δ^{13}C[J]. Can J Fish Aquat Sci, 2001, 58:1253-1260.

PAKTUNC D, FOSTER A, LAFLAMME G. Speciation and characterization of arsenic in Ketza River mine tailings using X-ray absorption spectroscopy[J]. Environ Sci Technol, 2003, 37:2067-2074.

PARSONS JG, ALDRICH MV, GARDEA-TORRESDEY JL. Environmental and biological applications of extended X-ray absorption fine structure (EXAFS) and X-ray absorption near edge structure (XANES) spectroscopies[J]. Appl Spectrosc Rev, 2002, 37:187-222.

PENG H, HU B, LIU Q, et al. Liquid chromatography combined with atomic and molecular mass spectrometry for speciation of arsenic in chicken liver[J]. Chromatogr A, 2014, 1370:40-49.

PEREIRA AA, VAN HATTUM B, DE BOER J, et al. Trace elements and carbon and nitrogen stable isotopes in organisms from a tropical coastal lagoon[J]. Arch Environ Contam Toxicol, 2010, 59:464-477.

PERERA PACT, SUNDARABARATHY TV, SIVANANTHAWERL T, et al. Arsenic and cadmium contamination in water, sediments and fish is a consequence of paddy cultivation: Evidence of river pollution in Sri Lanka[J].

Achievements in the Life Sciences, 2016, 10:144-160.

PETERSON BJ, FRY B. Stable isotopes in ecosystem studies[J]. Annu Rev Ecol Syst, 1987, 18:293-320.

PETRICK JS, AYALA-FIERRO F, CULLEN WR, et al. Monomethylarsonous acid (MMA^{III}) is more toxic than arsenite in Chang human hepatocytes[J]. Toxicol Appl Pharm, 2000, 163:203-207.

PHILIPS DL, NEWSOME SD, GREGG JW. Combining sources in stable isotope mixing models: alternative methods[J]. Oecologia, 2005, 144:520-527.

PIZARRO I, GÓMEZ M, CÁMARA C, et al. Arsenic speciation in environmental and biological samples extration and stability studies[J]. Analytica Chimica Acta, 2003, 495:85-98.

POLETTE LA, GARDEA-TORRESDEY JL, CHIANELLI RR, et al. XAS and microscopy studies of the uptake and bio-transformation of copper in Larrea tridentata (creosote bush)[J]. Microchem J, 2000, 65:227-236.

POST DM. The long and short of food-chain length[J]. Trends Ecol Evol, 2002a, 17:269-277.

POST DM. Using stable isotopes to estimate trophic position: Models, methods, and assumptions[J]. Ecology, 2002b, 83:703-718.

POST DM, Takimoto G. Proximate mechanisms for variation in food-chain length[J]. Oikos, 2007, 116:775-782.

PRICE RE, BREUER C, REEVES E, et al. Arsenic bioaccumulation and biotransformation in deep-sea hydrothermal vent organisms from the PACMANUS hydrothermal field, Manus Basin, Papua New Guinea[J]. Deep-Sea Res PT I, 2016, 117:95-106.

QUAGHEBEUR M, RENGEL Z, SMIRK M. Arsenic speciation in

terrestrial plant material using microwave-assisted extraction, ion chromatography and inductively coupled plasma mass spectrometry[J]. Anal At Spectrom, 2003, 18:128-134.

REVENGA JE, CAMPBELL LM, ARRIBÉRE MA, et al. Arsenic, cobalt and chromium food Web biodilution in a Patagonia mountain lake[J]. Ecotox Environ Safe, 2012, 81:1-10.

ROBINSON B, KIM N, MARCHETTI M, et al. Arsenic hyper accumulation by aquatic macrophytes in the Taupo Volcanic Zone, New Zealand[J]. Environ Exp Bot, 2006, 58:206-215.

ROIG N, SIERRA J, ORTIZ JD, et al. Integrated study of metal behavior in Mediterranean stream ecosystems: A case-study[J]. Hazard Mater, 2013, 263p:122-130.

ROMERO-FREIRE A, MARTÍN PEINADO FJ, DÍEZ ORTIZ M, et al. Influence of soil properties on the bioaccumulation and effects of arsenic in the earthworm *Eisenia andrei*[J]. Environ Sci Pollut R, 2015, 22:15016-15028.

RONKART SN, LAURENT V, CARBONNELLE P, et al. Speciation of five arsenic species (arsenite, arsenate, MMAA^V, DMAA^V and AsBet) in different kind of water by HPLC-ICP-MS[J]. Chemosphere, 2007, 66:738-745.

RUBIO-FRANCHINI I, LÓPEZ-HERNÁNDEZ M, RAMOS-ESPINOSA MG, et al. Bioaccumulation of metals arsenic, cadmium, and lead in zooplankton and fishes from the Tula River Watershed, Mexico[J]. Water Air Soil Pollut, 2016, 227:5-16.

RUIZ-CHANCHO MJ, LÓPEZ-SÁNCHEZ JF, SCHMEISSER E, et al. Arsenic speciation in plants growing in arsenic-contaminated sites[J]. Chemosphere, 2008, 71:1522-1530.

Ruttens A, Blanpain AC, Temmerman LD, et al. Arsenic speciation in food in Belgium: Part 1: Fish, molluscs and crustaceans[J]. Geochem Explor, 2012, 121:55-61.

SADEE B, FOULKES ME, HILL SJ. Coupled techniques for arsenic speciation in food and drinking water: a review[J]. Anal At Spectrom, 2015, 30:102-118.

SAIGO M, ZILLI FL, MARCHESE MR, et al. Trophic level, food chain length and omnivory in the Paraná River: a food web model approach in a floodplain river system[J]. Ecol Res, 2015, 30:843-852.

SANTOS CMM, NUNES MAG, BARBOSA IS, et al. Evaluation of microwave and ultrasound extraction procedures for arsenic speciation in bivalve mollusks by liquid chromatography‐inductively coupled plasma-mass spectrometry[J]. Spectrochimica Acta Part B: Atomic Spectroscopy, 2013, 86:108-114.

SCHAEFFER R, FRANCESCONI KA, KIENZL N, et al. Arsenic speciation in freshwater organisms from the river Danube in Hungary[J]. Talanta, 2006, 69:856-865

SHAH AQ, KAZI TG, ARAIN MB, et al. Hazardous impact of arsenic on tissues of same fish species collected from two ecosystem[J]. Hazard Mate, 2009a, 167:511-515.

SHARMA VK, SOHN M. Aquatic arsenic: toxicity, speciation, transformations, and remediation[J]. Environ Int, 2009, 4:743-759.

SHEN S, LI XF, CULLEN WR, et al. Arsenic binding to proteins[J]. Chem Rev, 2013, 113:7769-7792.

SINGH NK, RAGHUBANSHI AS, UPADHYAY AK, et al. Arsenic and

other heavy metal accumulation in plants and algae growing naturally in contaminated area of West Bengal, India[J]. Ecotox Environ Safe, 2016, 130:224-233.

ŠLEJKOVEC Z, BAJC Z, DOGANOC DZ. Arsenic speciation patterns in freshwater fish[J]. Talanta, 2004, 62:931-936.

SMEDLEY PL, KINNIBURGH DG. A review of the source, behaviour and distribution of arsenic in natural waters[J]. Appl Geochem, 2002, 17, 517-568.

SMITH PG, KOCH I, GORDON RA, et al. X-ray absorption near-edge structure analysis of arsenic species for application to biological environmental samples[J]. Environ Sci Technol, 2005, 39:248-254.

SMITH PG, KOCH I, REIMER KJ. An investigation of arsenic compounds in fur and feathers using X-ray absorption spectroscopy speciation and imaging[J]. Sci Total Environ, 2008, 390:198-204.

SOHRIN Y, MATSUI M, KAWASHIMA M, et al. Arsenic biogeochemistry affected by eutrophication in lake Biwa, Japan[J]. Environ Sci Technol, 1997, 31:2712-2720.

STYBLO M, RAZO LMD, VEGA L, et al. Comparative toxicity of trivalent and pentavalent inorganic and methylated arsenicals in rat and human cells[J]. Arch Toxicol, 2000,74:289-299.

SUHENDRAYATNA, OHKI A, MAEDA S. Biotransformation of arsenite in freshwater food chain models[J]. Appl Organometal Chem, 2001a,15:277-284.

SUHENDRAYATNA, OHKI A, NAKAJIMA T, et al. Metabolism and organ distribution of arsenic in the freshwater fish *Tilapia mossambica*[J]. Appl Organometal Chem, 2001b,15:566-571.

TANG JW, LIAO YP, YANG ZH, et al. Characterization of arsenic serious-contaminated soils from Shimen realgar mine area, the Asian largest realgar deposit in China[J]. Soil Sediment, 2016, 16,1519-1528.

THOMAS P, FINNIE JK, WILLIAMS JG. Feasibility of identification and monitoring of arsenic species in soil and sediment samples by coupled high-performance liquid chromatography-inductively coupled plasma mass spectrometry[J]. Anal Ato Spectrom, 1997, 12:1367-1372.

TOMY GT, PLESKACH K, FERGUSON SH, et al. Trophodynamics of some PFCs and BFRs in a western Canadian Arctic marine food web[J]. Environ Sci Technol, 2009, 43:4076-4081.

TRIPATHI RD, SRIVASTAVA S, MISHRA S, et al. Arsenic hazards:strategies tolerance and remediation by plants[J]. Trends Biotechnol, 2007, 25:158-165.

TURPEINEN R, PANTSAR-KALLIO M, HÄGGBLOM M, et al. Influence of microbes on the mobilization, toxicity and biomethylation of arsenic in soil[J]. Sci Total Environ,1999, 236:173-180.

VANDER ZANDEN MJ, CABANA G, RASMUSSEN JB. Comparing trophic position of freshwater fish caculated using stable nitrogen isotope ratios (δ^{15}N) and literature dietary data[J]. Can J Fish Aquat Sci, 1997,54:1142-1158.

VANDER ZANDEN MJ, CASSELMAN JM, RASMUSSEN JB. Stable isotope evidence for the food web consequences of species invasion in lakes[J]. Nature, 1999,401:464-467.

VANDER ZANDEN MJ, RASMUSSEN JB. Primary consumer δ^{13}C and δ^{15}N and the trophic position of aquatic consumers[J]. Ecology, 1999,80:1395-1404.

VANDER ZANDEN MJ, RASMUSSEN JB. Variation in the δ^{15}N and

$\delta^{13}C$ trophic fractionation: Implications for aquatic food web studies[J]. Limnol Oceanogr, 2001,46:2061-2066.

VAN GESTEL CAM, KOOLHAAS JE, HAMERS T, et al. Effects of metal pollution on earthworm communities in a contaminated floodplain area: Linking biomarker, community and functional responses[J]. Environ Pollut, 2009, 157:895–903.

VIZZINI S, COSTA V, TRAMATI C, et al. Trophic transfer of trace elements in an isotopically constructed food chain from a semi-enclosed marine coastal area (Stagnone di Marsala, Sicily, Mediterranean)[J]. Arch Environ Contam Tox, 2013,65:642-653.

WAN X, LEI M, CHEN T, et al. Micro-distribution of arsenic species in tissues of hyperaccumulator *Pteris vittata* L.[J]. Chemosphere, 2017, 166:389-399.

WANG S, LI B, ZHANG M, et al. Bioaccumulation and trophic transfer of mercury in a food web from a large, shallow, hypereutrophic lake (Lake Taihu) in China[J]. Environ Sci Pollut Res, 2012, 19:2820-2831.

WANG Y, WANG S, XU P, et al. Review of arsenic speciation, toxicity and metabolism in microalgae[J]. Rev Environ Sci Bio/Technol, 2015,14:427-451.

WATANABE K, MONAGHAN MT, TAKEMON Y, et al. Biodilution of heavy metals in a stream macroinvertebrate food web: Evidence from stable isotope analysis[J]. Sci Total Environ, 2008, 394:57-67

WEI CY, GE ZF, CHU W, et al. Speciation of antimony and arsenic in the soils and plants in an old antimony mine[J]. Environ Exp Bot, 2015, 109:31-39.

WELNA M, SZYMCZYCHA-MADEJA A, POHL P. Comparison of strategies for sample preparation prior to spectrometric measurements for determination and speciation of arsenic in rice[J]. Trac Trend Anal Chem, 2015, 65:122-136.

WHALEY-MARTIN K, KOCH I, MORIARTY M, et al. Arsenic speciation in blue mussels (*Mytilus edulis*) along a highly contaminated arsenic gradient[J]. Environ Sci Technol, 2012, 46:3110-3118.

WILLIAMS G, WEST JM, KOCH I, et al. Arsenic speciation in the freshwater crayfish, *Cherax destructor* Clark [J]. Sci Total Environ, 2009, 407:2650-2658.

YADAV A, RAM A, MAJITHIYA D, et al. Effect of heavy metal on the carbon and nitrogen ratio in Avicennia marina from polluted and unpolluted regions[J]. Mar Pollut Bull, 2015, 101:359-365.

YANG F, ZHANG N, WEI C, et al. Arsenic speciation in organisms from two large shallow freshwater lakes in China[J]. Bull Environ Contam Toxicol, 2017, 98:226-233.

YAO L, HUANG L, HE Z, et al. Delivery of roxarsone via chicken diet→chicken→chicken manure→soil→rice plant[J]. Sci Total Environ, 2016, 566-567:1152-1158.

YOSHIOKA T, WADA E, HAYASHI H. Stable isotope study on seasonal food web dynamics in a eutrophic lake[J]. Ecology, 1994, 75:835-846.

ZENG XB, SU SM, FENG QF, et al. Arsenic speciation transformation and arsenite influx and efflux across the cell membrane of fungi investigated using HPLC‐HG‐AFS and in-situ XANES[J]. Chemosphere, 2015, 119:1163-1168.

ZHANG W, CAI Y, TU C, et al. Arsenic speciation and distribution in an arsenic hyperaccumulating plant[J]. Sci Total Environ, 2002, 300:167-177.

ZHANG W, WANG WX. Large-scale spatial and interspecies differences in trace elements and stable isotopes in marine wild fish from Chinese waters[J]. Hazard Mater, 2012, 215-216:65-74.

ZHANG N, WEI C, YANG L. Occurrence of arsenic in two large shallow freshwater lakes in China and a comparison to other lakes around the world[J]. Microchem J, 2013,110:169-177.

ZHAO FJ, MA JF, MEHARG AA, et al. Arsenic uptake and metabolism in plants[J]. New Phytol, 2009,181:777-794.

ZHAO L, XU Y, HOU H, et al. Source identification and health risk assessment of metals in urban soils around the Tanggu chemical industrial district, Tianjin, China[J]. Sci Total Environ, 2013, 468-469C:654-662.

ZHENG J, HINTELMANN, DIMOCK B, et al. Speciation of arsenic in water, sediment, and plants of the Moira watershed, Canada, using HPLC coupled to high resolution ICP-MS[J]. Anal Bioanal Chem, 2003, 377:14-24.

ZHENG J, HINTELMANN H. HPLC-ICP-MS for a comparative study on the extraction approaches for arsenic speciation in terrestrial plant, *Ceratophyllum demersum*[J]. Radioanal Nucl Ch, 2009, 280:171-179.

ZHU X, WANG R, LU X, et al. Secondary minerals of weathered orpiment-realgar-bearing tailings in Shimen carbonate-type realgar mine, Changde, Central China[J]. Miner Petrol, 2015,109:1-15.

第 2 章

砷矿区概况及环境和生物样品砷形态和砷含量分析

2.1 石门雄黄矿概述

2.1.1 石门雄黄矿自然地理位置概况

湖南省石门县位于湖南省西北部，地处湖南和湖北交界处，古称"湘西北门户"，也是湖南省西北部铁路交通枢纽、电力能源中心、水泥建材基地，还是驰名中外的柑橘、茶叶之乡。石门县东连澧县、临澧县，南接慈利县、桃源县，西抵桑植县、鹤峰县，北毗五峰土家族自治县、松滋市。石门县隶属于湖南省常德市，总面积约 3973 km^2。地势自西向东南倾斜，海拔 42.5～2098.7 m，全县平均海拔为 500 m 左右。气候属于典型的亚热带大陆性季风气候，年平均气温为 16.7 ℃，年平均降水量为 1540 mm。纵横石门县的河流沟溪有 236 条，发源或流经石门县的有澧水、溇水、沱水、潨水、道水、黄水、涔水、沱水、澧水、道水，自西向东，分别贯穿石门县北部、中部、南部，入松滋市、临澧县，而后汇入洞庭湖。溇水从西北往东南纵贯，经泥沙、磨岗隘、袁公渡、新关等乡镇，至三江口注入澧水，长约 165 km。

近年来，石门县农业逐步向现代农业、生态农业迈进，初步形成以柑橘、马头羊、茶叶、高山蔬菜、干水果、烤烟为主的支柱经济产品和以粮食、生猪、家禽为主的传统产业产品。湖南省石门县是全国著名的传统名茶产地，是"中国名茶之乡""中国茶禅之乡""全国绿茶出口基地县""全国茶叶优势重点县""全国三绿工程茶叶示范县""湖南省有机茶第一县"（刘秋华 等，2015）。同时，石门县不仅是湖南第一柑橘大县，同时也是"中国柑橘之乡""中国早熟蜜橘第一县"。每年 9 月，全县日销柑橘在 2500 t 以上。石门县也是"湖南省旅游强县"。此外，石门县是湖南省矿产资源大县，有储量居世界之冠的雄磺矿，居亚洲之冠的矽砂矿、磷矿。石门县的矿产资源高达 30 多种，以非金属、固体燃料矿产为主，雄黄、矽砂、煤炭、石膏、石煤、磷矿等各类矿产的矿床、矿点有 320 多处（江开国，2000）。

2.1.2 石门雄黄矿地质及开采历史概况

石门雄黄矿位于湖南省石门县白云乡鹤山村，距石门县西北 41 km，与慈利县交界。石门县四面环山，中布二水两岭。西南屏石慈界山，海拔 919.5 m；西北为白羊山，海拔 714 m；东北为五峰山，海拔 701 m；东面分水岭较低，海拔 322.1 m。黄水溪从域内流过，注入澧水分支溇水。从地质角度而言，矿区构成一个近东西走向的长轴背斜，东西长达 10 余千米，背斜的轴部是寒武系的石灰岩，两翼由奥陶系石灰岩、志留系砂页岩组成。在背斜的轴部有一条东西走向的断层结构，背斜西段还有一条南北走向的横断层将背斜切割开。两条断层的交叉部位在成矿前即存在着古岩溶溶洞，因此雄黄矿既沿着断层发育，形成角砾状矿石，也沿着溶洞发育形成雄黄矿晶簇（黄镜友，1995）。

该矿是我国也是世界上最古老的矿山之一，以蕴含丰富的雄黄资源而闻名世界，为亚洲最大的雄黄矿，有 1500 多年开采历史。石门雄黄矿作为单独的砷矿床，平均含砷量 75%，矿石储量 730000 t（宣之强，1998）。1958 年后开始大规模开采和冶炼，年选矿 15000 t，生产原矿石约 3000 t（夏晶 等，2012）。经多年采挖，300000 t 的雄黄储量已消耗殆尽，资源濒临枯竭。1978 年，炼砷厂被关闭，但仍有硫酸、磷肥和水泥厂在生产。2001 年，石门雄黄矿因资源枯竭，经国土资源部批准正式闭矿；2011 年，因污染严重，区内硫酸、磷肥和水泥厂被依法关停。2008 年，距矿区下游 11km 的皂市水库建成使用，主要用于防洪和发电。由于开采冶炼排放的废气、废渣对当地大气、水体、土壤、牲畜、蔬菜和农作物等造成了长期、严重的污染。2012 年 10 月，《石门雄黄矿区重金属污染"十二五"综合防治实施方案》开始实施。雄黄矿污染范围在石门县内主要是磺厂社区、鹤山村、望羊桥村，涉及人口 4500 多人。石门雄黄矿是亚洲最大的单砷矿，是土壤砷高背景区和矿冶活动共同导致重金属严重污染的典型代表。矿区冶炼点及周边河流和水库如图 2.1 所示。

(a) 原砷矿冶炼点　　　　　(b) 上游河流（田间）

(c) 上游河流（村庄）　　　　(d) 皂市水库

图 2.1　矿区冶炼点及周边河流和水库

The mining sites 和 river around the realgar mine area

2.2 石门雄黄矿环境及生物样品实验方法

2.2.1 环境和生物样品采集及预处理

2.2.1.1 水样采集及预处理

水样选取表层水（0～20 cm），现场用 YSI6600V2 型水质多参数分析仪检测水温、pH 值、氧化还原电位、溶解氧等水体理化性质。每个点位采集三份 500 mL 水装入纯净水瓶之中，其中两份取 100 mL 作为原水测总砷含量（每个样品滴加 2～3 滴硝酸），另取 100 mL 过滤（0.45 μm 膜），转移于采样瓶中低温冷藏，作为过滤水用于总砷和砷形态分析。其中，水体总氮（TN）、总磷（TP）和叶绿素 a（Chl-a）的测定由中国科学院南京地理与湖泊研究所完成。

2.2.1.2　土壤和表层沉积物采集及预处理

水样采集的同时，用彼得森采样器采集表层（30 cm 左右）沉积物，置于塑料自封袋内，带回实验室冷藏 24 h，取上层混合液至 50 mL 离心管中，离心取间隙水，过 0.45 μm 膜，用于总砷和砷形态测试。沉积物直接风干，磨碎至粉末状备用。

土壤采样点主要位于河流边上的农田、尾矿及河床等，每个采样点取 0~20 cm 的表层土壤，去除表面碎石及其他覆盖物，使用梅花 5 点法在每个采样点采集 1~2 kg 混合土壤样品。样品除去石块、植物根系等杂物，自然风干，磨碎分别过 10 目、200 目尼龙筛，编号保存备用。

2.2.1.3　水生生物采集及预处理

生物样品采集后现场称重，测量体长，鉴定物种，解剖分离。所有生物样品先用自来水清洗，再用纯净水洗净，编号转入自封袋，在野外用冰袋冷藏，放入-20℃冰箱，后用保温箱低温运回实验室立即冷冻保存。所有样品用冷冻干燥仪（Techconp FD-3-85-MP）干燥 72 h 后，在液氮下采用全自动样品快速研磨仪粉碎至粉末状后冷藏以备分析。

2.2.1.4　陆生生物采集及预处理

选取一定大小的样方，采集蚯蚓、凋落物、地表植物，同时采集栖息点的土壤样品。将采集的蚯蚓洗净表面泥土后，放入铺有湿润中性滤纸的托盘中，用事先已扎好通气孔的塑料薄膜封口，室温下清肠吐泥 48 h，期间每隔 12 h 更换滤纸一次，彻底冰冻后用手术刀解剖，用去离子水洗净，尽量保证蚯蚓体内的异物充分排出。在液氮下采用全自动样品快速研磨仪粉碎至粉末状后冷藏以备分析。

地表植物的采集分为根、茎、叶等组织。凋落物、地表植物分别置于纸质信封中,用去离子水清洗干净。前人通过比较发现,植物在新鲜、冻干及烘干前处理条件下,其As(Ⅲ)与As(Ⅴ)的比例基本一致,说明烘干基本不会影响植物中的As(Ⅲ)氧化为As(Ⅴ)(Jedynak et al., 2010)。因此,本研究对地表植物、凋落物进行烘干(70℃)后采用全自动样品快速研磨仪研磨至粉末状。

2.2.2 石门雄黄矿环境和生物样品分析

2.2.2.1 环境和生物样品多指标测试所需实验试剂和仪器

所用试剂有:HNO_3(优级纯,70%),$HClO_4$(优级纯,71%),HCl(优级纯),H_2O_2(优级纯,30%),KBH_4(优级纯),KOH(优级纯),硫脲(优级纯),抗坏血酸(优级纯),$FeSO_4 \cdot 7H_2O$(优级纯),$K_2Cr_2O_7$(优级纯),H_2SO_4(优级纯),$(NH_4)_2HPO_4$(优级纯),$(NH_4)_2CO_3$(分析纯,71%),氨水(优级纯),甲酸(优级纯),甲醇(HPLC 纯)等。实验用水均为超纯水(18.2 $M\Omega \cdot cm$),由 Milli-Q 超纯水器制得。所有玻璃器皿均用自来水冲洗干净,晾干后用10%的HNO_3浸泡24~48 h,再用自来水冲洗5遍、蒸馏水淋洗3遍,放烤箱烤干后备用。

2.2.2.2 环境和生物样品稳定同位素分析方法

生物样品冻干之后,所有鱼类样品取其背部肌肉,青蛙、鸟类、螃蟹等取其后腿肌肉,用于同位素分析(Watanabe et al., 2008; Vizzini et al., 2013)。全部待测样品在60℃烘干至恒重,用研钵研磨成粉末状后放入干燥器中保存待测。分析所用仪器为 MAT 253 稳定同位素质谱仪,为德国 Finnigan 公司生产的最新型稳定同位素比值质谱仪,具高灵敏度和高线性特征。同位素分析在中国科学院地理科学与资源研究所理化中心完成。

由于样品与标准参照物之间比率差异较小,所以稳定同位素丰度表示为样品与标准之间的偏差的千分数:

$$\delta^{13}C_{样品丰度}=[(^{13}C_{含量}/^{12}C_{样品含量})/(^{13}C_{含量}/^{12}C_{标准含量})-1]\times1000$$

$$\delta^{15}N_{样品丰度}=[(^{15}N_{含量}/^{14}N_{样品含量})/(^{15}N_{含量}/^{14}N_{标准含量})-1]\times1000$$

碳稳定同位素测定的标准物质是美洲拟箭石化石(PDB),其$\delta^{13}C$丰度被定义为0‰。氮稳定同位素测定的标准物质是纯化的大气中的N_2,其$\delta^{15}N$丰度被定义为0‰。测试碳、氮同位素时为保持实验结果的准确性和仪器的稳定性,每测试5个样品就插测1个标准样品,所有样品均重复测试2次。采用尿素样品量:$\delta^{13}C$:−43.77‰±0.35‰;$\delta^{15}N$:−1.21‰±0.06‰。

2.2.2.3 石门雄黄矿土壤理化性质实验分析方法

1)土壤含水率实验分析方法

称取采集的新鲜土壤样品(>1 mm风干土)20 g左右,置于事先称好重量的干净铝盒中。之后将装有土壤的铝盒放入烘箱,在100℃左右(温度过高,有机质容易碳化散逸)温度下烘至恒重。取出铝盒放在干燥器中冷却半小时,立即称重。结果计算:

$$W=(W_2-W_3)/(W_3-W_1)\times100$$

式中:

W为含水率(%);

W_1为铝盒重量(g);

W_2为铝盒+样品重量(g);

W_3为铝盒+烘干后土重(g)。

2) 土壤pH值实验分析方法

称取10 g（±0.01 g）样品（过2 mm筛），置于50 mL的高型烧杯或其他适宜的容器中，并加入除去CO_2的超纯水25 mL，用玻璃棒搅拌5 min，然后静止1～2 h。依照雷磁pH值计仪器说明书，至少使用两种pH值标准缓冲溶液进行pH值校正。测量pH值时，在搅拌的条件下或事前充分摇动试样溶液后，将pH值计玻璃电极插入试样溶液中，待读数稳定后读取pH值。

3) 土壤有机质实验分析方法

准确称量0.15～0.25 g的样品（视样品中含有机质的多少而定，含量越高，取样量越少）倒入锥形瓶中。向锥形瓶中准确滴加1N（0.167 mol/L）的$K_2Cr_2O_7$标准溶液5 mL，然后再加入6 mL左右的H_2SO_4，振荡使之充分反应。盖上表面皿，电热板显示温度范围为210～220℃，将电热板台面温度控制在170～180℃。自锥形瓶内溶液开始沸腾并出现大量气泡时开始计时，保持溶液沸腾5±0.5 min。消煮时间准确与否对分析结果影响较大，严格控制消煮时间。将锥形瓶取出，冷却至室温，加20 mL左右去离子水稀释溶液酸度。加几滴指示剂（配好的邻菲罗啉指示剂），用0.5 mol/L左右的$FeSO_4$溶液滴定至溶液变成棕红色（先是深绿色，之后是蓝绿色，之后为棕红色），滴定时在白炽灯产生的强光下观察变色反应。最后，记录消耗的$FeSO_4$溶液的体积。按同样的方法处理空白实验，记录消耗的$FeSO_4$溶液的体积。同批次做2个空白实验，密切注意滴定终点。

土壤有机质计算公式：

$$\text{土壤有机质 (g/kg)} = 1.724 \times \left\{ \frac{\frac{c \times 5}{v_0} \times (v_0 - v) \times 10^{-3} \times 3.0 \times 1.1}{m} \right\} \times 1000$$

式中：

c 为 0.8 mol/L（1/6 $K_2Cr_2O_7$）标准溶液的浓度；

5 表示加入 $K_2Cr_2O_7$ 标准溶液的体积（mL）；

v_0 为空白实验滴定用去 $FeSO_4$ 溶液的体积（mL）；

v 为样品滴定用去 $FeSO_4$ 溶液的体积（mL）；

10^{-3} 表示将 mL 换算为 L；

3.0 表示 $\frac{1}{4}$ 碳原子的摩尔质量（g/mol）；

1.1 表示氧化校正系数；

m 为风干样品质量（g）；

1.724 表示土壤有机碳转换成土壤有机质的平均换算系数。

2.2.3 石门雄黄矿环境和生物样品总砷量测定方法

2.2.3.1 水样品总砷量测定方法

过滤水样品可直接上机测总砷量，具体方法如下：取 5 mL 待测样品于 10 mL 具塞刻度试管中，加入 5 mL 还原剂（10% HCl、2%硫脲和 2%抗坏血酸），摇匀后静置 1 h。选用氢化物发生—原子荧光光谱仪，其中载流：5% HCl，还原剂：2% KBH_4 + 0.5% KOH，正确连接仪器管路，稳定仪器（仪器预热时间半小时以上），采用标准曲线法测总砷量。

原水中砷含量的测试需先消解，过程如下：将待测样品事先摇匀，用移液枪准确移取 20 mL 原水至锥形瓶中，加入 10 mL HNO_3 和 $HClO_4$ 混合液（体积比 HNO_3 : $HClO_4$ = 8 : 1）。盖上表面皿，放于电热板上加热（180℃）

消解至白色浓烟冒尽。取出冷却后，用超纯水少量多次冲洗烧杯至 25 mL，置于比色管或具塞刻度试管中，摇匀静置，按照检测过滤水样的方法用 HG-AFS 测总砷量。

2.2.3.2 土壤及沉积物样品总砷量测定方法

称取约 50 mg 土壤或沉积物样品，放入 100 mL 玻璃烧杯中，加入 10 mL HNO_3 和 $HClO_4$ 混合液（体积比 HNO_3：$HClO_4$ = 5：1），盖上表面皿，室温静置过夜。然后置于电热板上加热（180℃）消解至白色浓烟冒尽，残余样品呈（灰）白色，继续加热至 0.5 mL 左右。若消解不完全，继续加酸进一步消解。消解完全后，取出冷却，用超纯水定容至 25 mL，置于比色管或具塞刻度试管中，摇匀静置，按照检测过滤水样的方法用 HG-AFS 测总砷量。

2.2.3.3 动物样品总砷量测定方法

称取约 100~300 mg 动物样品，放入 25 mL 比色管或具塞刻度试管中，加入 3.5 mL 的 HNO_3，盖上带孔玻璃塞，室温静置过夜。次日，样品基本溶于 HNO_3 中，用电热板缓慢升温至 125℃，加热 1~2 h。调节电热板温度至 140℃，使得溶液轻度沸腾。若消解不完全，继续加酸进一步消解。继续蒸发浓缩至溶液剩约 1 mL 时，取下静置冷却，加入 1.5 mL 的 H_2O_2，在电热板上继续加热，待剩余溶液至 0.5 mL 左右，取下冷却，用超纯水定容至 25 mL，摇匀静置，按照检测过滤水样的方法用 ICP-MS 测总砷量。

2.2.3.4 植物样品总砷量测定方法

称取约 500 mg 凋落物、植物等样品，放入 100 mL 玻璃烧杯中，加入 10 mL 的 HNO_3 和 $HClO_4$ 混合液（体积比 HNO_3：$HClO_4$ = 9：1），盖上表

面皿，室温静置过夜。次日，将溶液置于电热板上加热（180℃）消解至白色浓烟冒尽，残余样品呈现（灰）白色，继续加热溶液至 0.5 mL 左右。若消解不完全，继续加酸进一步消解。取出冷却后，用超纯水定容至 25 mL，置于比色管或具塞刻度试管中，摇匀静置，按照检测过滤水样的方法用 ICP-MS 测总砷量。

2.2.4 石门雄黄矿环境和生物样品砷形态分析方法

2.2.4.1 水样品砷形态分析方法

水样品中的砷形态主要采用 HPLC-(UV)-HG-AFS 测试（仪器相关参数配置见表 2-1），包括过滤水和间隙水，直接上机测试。

表 2.1 HPLC-(UV)-HG-AFS 仪器相关参数配置

The experimental parameters for HPLC-(UV)-HG-AFS

检测技术		HPLC-UV-HG-AFS
色谱参数	阴离子交换柱	Hamilton PRP-X100（250 mm×4.1 mm i.d., 10μm）
	四通切换阀阀位	UV
	流动相	5 mmol/L $(NH_4)_2HPO_4$（用氨水调节 pH 值：9.0）
	流速	1.0 mL/min
	进样体积	100 μL
氢化物发生参数	管道 1	还原剂：2%KBH_4+0.5%KOH
	管道 2	载流：7% HCl
	管道 3	间隔气：空气
	管道 4	载流：7% HCl
	元素灯	As（100 mA/45 mA）
AFS 参数	负高压	290 V
	灯电流	80 mA
	载气	氩气，400 mL/min
	屏蔽气	氩气，500 mL/min
	测定形态	As(III), DMA, MMA, As(V), AsB

2.2.4.2 动植物样品砷形态分析方法

本研究中动植物样品的砷提取过程参照文献[Ciardullo et al.，2010]。过程如下：称取 100~500 mg 的生物样品（主要根据生物样品的砷含量高低，高砷生物称取量约 100 mg，低砷生物称取量约 300~500 mg）置于 50 mL 离心管中，缓慢加入 15 mL 甲醇—水混合溶液（体积比 1∶1）。将离心管放置在水浴恒温震荡箱中，转速设置为 150 rpm，温度设置为 20 ℃。次日，将离心管取出，在室温下超声震荡 30 min，然后置于高速离心器中，转速 3000 rpm，离心 20 min。转移上清液至 25 mL 刻度试管中，冷藏。对剩余生物样品，继续添加 5 mL 甲醇—水混合溶液（体积比 1∶1），恒温震荡 1 h，再超声震荡 30 min，离心 20 min。整合两次提取的上清液，混合均匀。研究表明，一定量的甲醇会影响 ICP-MS 测试砷的检测信号及灵敏度（Guo et al.，2011；An et al.，2015，2017）。因此，为降低甲醇的影响，采用氮吹仪吹赶提取液中的甲醇，浓缩至 3 mL 左右，然后用超纯水定容至 10 mL。过 0.22 μm 滤膜，冷藏。用 HPLC-ICP-MS 联用系统检测主要砷形态（其工作参数见表 2.2）。

表 2.2 HPLC-ICP-MS 联用系统工作参数

Parameters of the HPLC-ICP-MS system

ICP-MS 参数
模式：Elan DRC-e
正向功率：1.45 kW
雾化器流速：1 L/min
载气流速：15 L/min
DRC 反应池：0.45 mL/min 的 O_2
Rpa：0.6
HPLC 参数
色谱柱：Hamilton PRP-X100，10 μm，4.1 mm×250 mm
色谱柱温度：40℃

续表

分离模式：isocratic	
流动相：30 mmol/L 的$(NH_4)_2CO_3$，pH=9.3（NH_4OH）	
流速：1.2 mL/min	
进样量：30 μL	
运行时间：40 min	
质谱参数	
检测信号：AsO m/z 91	
扫描模式：Peak Hopping	
重复次数：3	

2.2.4.3 X射线吸收光谱砷形态分析方法

砷的近边吸收谱在北京正负电子对撞机同步辐射实验室（BSRF）1W1B-XAFS 实验站进行测定。由于机时有限，只能对部分样品进行测试。对于砷含量较高的样品（砷含量>100 mg/kg），采用荧光探测器 Lyttle 荧光分离室采集荧光信号；而对于砷含量较低的样品（砷含量<10 mg/kg），选取同一类样品中含量相对较高的样品进行测试，在固探模式下采用 19 元高纯锗半导体阵列探测器采集荧光信号。本研究中的生物主要来自野外采集，其砷含量大多<10 mg/kg，尽管存在有机砷，但含量极低。同样，这也是线性组分分析（LCF）存在相对大误差的原因，其他研究中也有报道（Niazi et al.，2011；Kopittke et al.，2014；Foust Jr. et al.，2016）。

砷的参比物质选用液体的 As(V)、As(III)、DMA、AsB、AsC 和 As(III)-GSH 进行测试。其中 As(III)采用 $Na_2HAsO_4·7H_2O$（pH=5.5）配制，As(V)采用 $NaAsO_2$（pH=9.0）配制。As(III)-GSH 通过在 pH=5.5 的条件下用 10∶1（摩尔比）的谷胱甘肽与 $Na_2HAsO_4·7H_2O$ 合成（Pickering et al.，2000；George et al.，2009）。缓冲液选用 HEPES/NaOH+30%甘油。XAFS 实验站测定的砷的近边吸收光谱经过边前背景扣除、数据归一化、μ_0 拟合、k 空间转换及加权、快速傅里叶变换获得 XAFS 图谱。数据的预处理和傅里叶

变换通过 Athena 软件完成。获得的 X 射线吸收谱经过边前背景扣除和归一化处理后，选取能量为 11850eV-11900eV 的近边谱进行样品主要组分线性组成分析（Ravel et al.，2005）。

2.2.5　石门雄黄矿环境和生物样品中其他重金属含量测定方法

样品中其他重金属（Cd、Pb、Cr、Cu、Mn、Ni、Zn）的含量由中国科学院地理科学与地理资源所理化分析中心测定，选用 PerkinElmer 公司生产的电感耦合等离子体光谱仪 ICP-OES（仪器型号：Elan DRC-e）测定，对于含量低于 ICP-OES 检测限的元素，如 Cd 和 Pb，采用电感耦合等离子体质谱仪 ICP-MS（仪器型号：Optima 5300DV）测定。

2.2.5.1　水样品中其他重金属含量测定方法

水样品经过 0.45μm 膜过滤后可直接上机测定。

2.2.5.2　土壤样品中其他重金属含量测定方法

准确称量 0.0500g 样品，转移至 100 mL PTFE 烧杯中，滴入 1~2 滴超纯水，轻轻晃动 PTFE 烧杯，让所有样品聚在一起，盖上盖子。打开盖子，用玻璃移液管吸取 5mL HNO_3 至烧杯，用塑料移液管吸取 5 mL HF 至烧杯，用玻璃移液管吸取少量 $HClO_4$ 至烧杯，盖上盖子，轻微晃动烧杯，摇散烧杯底部样品，然后放到电热板上静置过夜。次日，打开电热板消解样品，电热板表面温度为 180℃，持续消解，大约每 1 h 揭开盖子一次，观察烧杯内样品消解情况。若烧杯内剩余液体体积约黄豆大小，液体呈透明或者浅黄绿透明色，且液体晶莹透亮，同时烧杯底部无任何残余样品时，样品消解完成。若烧杯内液体呈透明色，但仍有样品颗粒残余，则取下冷却后，加入少量 HF 和少量 HClO₄，放在电热板上继续消解。若烧杯内液体浑浊，则取下冷却后，按比例再次加入 HNO₃、HF、HClO₄，放在电热板上继续消解。消解彻底完成后，在烧杯中加入少量超纯水，放在电热板上加热 10～20 min（作用：将烧杯内壁上可能残余的样品溶解在水溶液内）。取下冷却后，用超纯水冲洗烧杯和盖子内壁，将消解液小心转移至 25 mL 玻璃试管中，用超纯水定容至 25 mL，加塞摇匀。定容后的样品，静置 3 h 后可直接上机测定。

2.2.5.3　动植物样品中其他重金属含量测定方法

准确称取 500 mg 凋落物、植物等样品，放入 100 mL 玻璃烧杯中，加入 10 mL HNO₃ 和 HClO₄ 混合液（9 mL HNO₃ 和 1 mL HClO₄），盖上表面皿，浸泡过夜。然后于电热板上加热（180℃）消解至白色浓烟冒尽，残余样品呈（灰）白色，继续加热至尽干。取出冷却后，用超纯水少量多次冲洗烧杯至 25 mL，置于比色管或具塞刻度试管中，摇匀静置，定容后的样品静置 3 小时后可直接上机测定。

2.3 石门雄黄矿环境和生物样品总砷和砷形态质量控制

为了保证结果的准确性和可靠性，我们以标准工作曲线、空白实验、标准物质的测定、平行样及样品加标回收实验，对实验数据进行质量控制。对出现异常值的样品重复进行消解和测定，避免其测定值受实验条件影响过大，最终数值通过综合分析几次实验结果得到。

2.3.1 石门雄黄矿环境和生物样品总砷质量控制

土壤、沉积物及动植物样品在消解过程中，在每批样品（30～45个）中设置3～5个标准样品，总砷分析质量控制结果如表2.3所示。

表 2.3 总砷分析质量控制结果

The quality control data of standard materials

标准物质	编号	参考值 (mg/kg)	n	测试值 (mg/kg)	回收率 (%)
土壤	GBW07453（GSS-24）	15.8±0.9	8	16.14±1.51	102.1
水系沉积物	GBW07308	2.4±0.6	5	2.50±0.37	104.2
水系沉积物	GBW07307	84±6	3	85.72±8.12	102.1
水系沉积物	GBW07304（GSD-4）	19.7±2.6	4	19.35±2.46	98.22
金枪鱼组织	BCR-627	4.8±0.3	11	4.65±0.09	96.87
扇贝	GBW10024	3.60±0.6	9	3.69±0.22	102.5
灌木枝叶	GBW07603（GSV-2）	1.25±0.15	9	1.21±0.04	96.80
茶叶	GBW08505	0.191±0.027	3	0.154±0.016	80.63

消解后在测总砷的样品（消解液）及过滤水样品中每 10 个样品里设置 1 个空白样品，并随机放置 2 μg/L、5 μg/L、10 μg/L 的砷模拟天然水砷溶液成分分析标准物质（GBW(E)080390），回收率为 86%~98%。另外，随机选取样品做加标回收及加标平行，回收率 79%~105%。

2.3.2 石门雄黄矿环境和生物样品砷形态质量控制

生物样品的砷形态分析过程比较复杂，为了正确认识砷形态提取方法的有效性和实验提取过程中不同砷形态的稳定性，采用样品加标回收和标准加标回收两种质量控制方法。不同砷形态标准溶液是从中国计量科学研究院购买的，有 GBW08667 砷酸根溶液标准物质——As(Ⅴ)，GBW08666 亚砷酸根溶液标准物质——As(Ⅲ)，GBW08668 一甲基砷溶液标准物质——MMA，GBW08669 二甲基砷溶液标准物质——DMA，GBW08670 砷甜菜碱溶液标准物质——AsB，GBW08671 砷胆碱溶液标准物质——AsC。6 种砷形态的色谱流出图如图 2.2 所示。砷形态标准溶液现配现用，且必须保存在-4℃黑暗条件下。

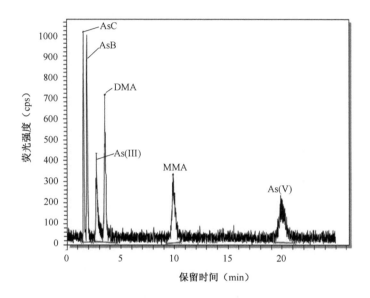

图 2.2　6 种砷形态的色谱流出图（4 μg/L）

HPLC chromatograms of the six standard arsenic species

提取和分析方法的质量保证与质量控制使用标准物质（SRM）BCR-627 金枪鱼组织（Institute for Reference Materials and Measurements，IRMM）进行检测。在 BCR-627 标准物质中，检测到 AsB 平均浓度为 3.69±0.06 mg/kg（参考值：3.90±0.22 mg/kg，回收率 91%，n=13）。DMA 平均浓度为 1.2±0.06 mg/kg（参考值：1.5±0.02 mg/kg，回收率 89%，n=13）。采用加标回收率方法确定其他砷形态的回收率如下：AsC，71%～115%；AsB，69%～107%；As(Ⅲ)，63%～129%；DMA，86%～128%；MMA，72%～117%；As(Ⅴ)，71%～123%。

2.4 本章小结

本章概述了石门雄黄矿的自然地理位置及矿区发展情况，对围绕石门雄黄矿及周边地区开展的野外样品采集及前处理过程进行了介绍，并详细介绍了不同样品进行室内分析（主要为总砷和砷形态测定）的操作流程，还介绍了质量控制以确保样品检测数据的准确性和可靠性。

参 考 文 献

黄镜友. 湖南境内沿倾向发育较深的脉状矿床的地质特征及规律[J]. 湖南地质, 1995, 14:44-47.

江开国. 加快石门矿产资源开发利用的战略思考[J]. 湖南地质, 2000, 19:111-116.

李忠义, 左涛, 戴芳群, 等. 运用稳定同位素技术研究长江口及南黄海水域春季拖网渔获物的营养级[J]. 中国水产科学, 2010, 17:103-109.

刘秋华, 王冰, 杨载田. 湖南省石门县名茶生产及其茶文化的可持续发展[J]. 农业考古, 2015, 5: 61-65.

夏晶, 曹帅, 吴赵云, 等. 药用雄黄的基源考证及实地调研[J]. 中华中医药杂志, 2012, 27:1543-1546.

宣之强. 中国砷矿资源概述[J]. 化工矿产地质, 1998, 20:205-211.

APOSHIAN HV, GURZAU EAN J, LEE J, LEE G, et al. Combined use of collision cell technique and methanol addition for the analysis of arsenic ina high-chloride-containing sample by ICP-MS[J]. Microchem J, 2015, 120:77-81.

AN J, LEE H, NAM K, et al. Effect of methanol addition on generation of isobaric polyatomic ions in the analysis of arsenic with ICP-MS[J]. Microchem J, 2017, 131:170-173.

CIARDULLO S, AURELI F, RAGGI A, et al. Arsenic speciation in freshwater fish: Focus on extraction and mass balance[J]. Talanta, 2010, 81:213-221.

FOUST JR RD, BAUER AM, COSTANZA-ROBINSON M, et al. Arsenic transfer and biotransformation in a fully characterized freshwater food web[J]. Coordin Chem Rev, 2016, 306:558-565.

GEORGE GN, PRINCE RC, SINGH SP, et al. Arsenic K-edge X-ray absorption spectroscopy of arsenic in seafood[J]. Mol Nutr Food Res, 2009, 53:552-557.

GUO W, HU S, LI X, et al. Use of ion-molecule reactions and methanol addition to improve arsenic determination in high chlorine food samples by DRC-ICP-MS[J]. Talanta, 2011, 84:887-894.

JEDYNAK L, KOWALSKA J, KOSSYKOWSKA M, et al. Studies on the uptake of difference arsenic forms and the influence of sample pretreatment on arsenic speciation in White mustard (*Sinapis alba*)[J]. Microchem J, 2010, 94:125-129.

KOPITTKE PT, DE JONGE MD, WANG P, et al. Laterally resolved speciation of arsenic in roots of wheat and rice using fluorescence-XANES imaging[J]. New Phytol, 2014, 201:1251-1262.

NIAZI NK, SINGH B, SHAH P. Arsenic speciation and phytoavailability in contaminated soils using a sequential extraction procedure and XANES spectroscopy[J]. Environ Sci Technol, 2011, 45:7135-7142.

PICKERING IJ, PRINCE RC, GEORGE MJ, et al. Reduction and coordination of arsenic in Indian mustard[J]. Plant Physiol, 2000, 122:1171-1177.

RAVEL B, NEWVILLE M. Athena, artemis, hephaestus: Data analysis for X-ray absorption spectroscopy using IFEFFIT[J]. J Synchrotron Radiat, 2005, 12, 537-541.

VIZZINI S, COSTA V, TRAMATI C, et al. Trophic transfer of trace elements in an isotopically constructed food chain from a semi-enclosed marine coastal area (Stagnone di Marsala, Sicily, Mediterranean)[J]. Arch Environ Contam Tox, 2013,65:642-653.

WATANABE K, MONAGHAN MT, TAKEMON Y, et al. Biodilution of heavy metals in a stream macroinvertebrate food web: Evidence from stable isotope analysis[J]. Sci Total Environ, 2008, 394:57-67.

第 3 章

砷矿区环境介质中砷的分布特征

第3章 砷矿区环境介质中砷的分布特征

3.1 砷矿区环境介质中砷的分布的研究意义

砷作为一种有毒的重金属，在环境中普遍存在。砷可通过饮水或沿着食物链进入人体，导致皮肤、肺、肾、膀胱等器官的病变，乃至诱发癌症（Mandal and Suzuki，2002；Hughes et al.，2011）。为降低砷的健康风险，1993年，WHO 将生活饮用水砷标准从 50 μg/L 改为 10 μg/L。我国也于 2007 年正式将生活饮用水砷标准修订为 10 μg/L。自然释放加上工业化、城市化后不断增强的人类活动，使得大量的砷进入环境中（Mandal and Suzuki，2002）。砷一旦进入土壤中，很难被去除，长期积累在土壤中的砷很容易通过农作物转移到人体及其他生物体中，从而造成潜在危害（黄益宗 等，2013）。土壤中的砷还可通过侵蚀和淋溶作用进入水体。水体中的砷主要为无机砷，很容易随水流动，或呈溶解态随水流动，或吸附于水合金属氧化物上悬浮于水中，或通过沉淀作用进入底泥中（杨芬 等，2015）。底泥中的砷可在氧化—还原、溶解和解吸等作用下，重新释放进入水体（Wang and Chapman，1999；Mdegela et al.，2009）。通过各种途径输送入水体的砷，绝大部分迅速进入沉积物中，因此沉积物可以认为是砷在水环境中迁移转化的最终归宿（Brinton，2001）。

  王振刚等人（1999）于 1994 年对石门雄黄矿地区的环境污染状况进行了现场调查，发现土壤中的砷含量为 84~296 mg/kg，河水中的砷含量为 0.5~15.8 mg/L。2001 年石门雄黄矿正式闭矿，2012 年《石门雄黄矿重金属污染"十二五"综合防治方案》实施，开始清理历史遗留砒渣，对周边污染土壤进行治理。但是 Tang 等人（2016）研究发现，石门雄黄矿中心区域的土壤中的砷含量最高仍可达到 5240 mg/kg；Zhu 等人（2015）研究发现，尾矿和冶炼区水体中的砷含量高达 40.10 mg/L。因此，全面查清石门雄黄矿的污染现状，对石门雄黄矿区水体、沉积物和土壤中砷的分布特征进行分析，对于治理该矿区环境中的砷污染具有重要意义。

3.2 砷矿区环境介质采集、总砷及砷形态分析方法

作者团队分别于 2015 年 5 月和 2016 年 6 月，从原砷矿中心区上游开始，顺黄水溪而下，进入溧水，共采集了 46 个点位的表层水样，27 个点位的底泥样品，以及 43 个点位的土壤样品。考虑到 2016 年采样期间出现连续暴雨，本研究将每个点位的表层土壤划分为三个深度（0~15 cm，15~30 cm 和 30~45 cm）进行采集，并混合。样品的采集、预处理及实验测试方法见第 2 章。

根据采样点位与原砷矿中心区的距离将采样区域划分为 7 个区域，分别为：区域 1（上游 1~8 km），区域 2（上游 0~1 km），区域 3（下游 0~1 km），区域 4（下游 1~2 km），区域 5（下游 2~3.5 km），区域 6（下游 3.5~5.5 km），区域 7（下游 5.5~7 km，即皂市水库所在区域）。

3.3 砷矿区环境介质中砷的分布特征

3.3.1 水体中砷的分布特征

3.3.1.1 水体理化性质

水体的水质参数如表 3.1 所示,不同区域的水质参数差异极为显著。矿区上游(区域 1)水体为弱酸性,pH=6.76;而在区域 2~区域 7,水体均为碱性,8.23≤pH≤8.84。其他参数的范围为:水温(20.6~26.5℃),ORP(138~210 mv),DO(8.95~10.40 mg/L),TDS(60~156 mg/L),Ec(93~240 μs/cm)。

表 3.1　水体的水质参数（平均值）

The water quality parameters in each sampling area

	区域1	区域2	区域3	区域4	区域5	区域6	区域7
样品数量	9	6	4	4	6	6	11
水温（℃）	20.6	24.7	26.5	25.2	25.8	23.8	23.9
ORP（mv）	210	171	153	144	138	149	153
pH 值	6.76	8.23	8.43	8.56	8.73	8.79	8.84
DO（mg/L）	9.30	9.20	8.95	9.17	9.09	10.40	10.20
Ec（μs/cm）	93	102	131	240	156	145	128
TDS（mg/L）	60	66	85	156	101	94	83
TN（mg/L）	1.98	2.31	2.34	2.14	2.09	1.71	1.41
TP（mg/L）	0.03	0.13	0.81	0.65	0.58	0.08	0.02
Chl-a（μg/L）	<dl	0.28	<dl	<dl	1.20	15.7	14.8

注：<dl 表示低于检测限。

不同点位的营养化参数（TP、TN 和 Chl-a）的差异也很显著。TP 的基础值范围为 0.02~0.81 mg/L，TN 的基础值范围为 1.41~2.34 mg/L，Chl-a 的基础值范围为未检测出到 15.7 μg/L。TN（总氮）、TP（总磷）水平参照 GB3838-2002，Chl-a（叶绿素 a）水平参照 GHZB1-1999。根据水体的富营养化水平标准，研究区域水系均呈现明显的中富营养化水平；特别是区域 3 和区域 4（砷矿中心冶炼区和尾矿区），其总磷、总氮含量最高，可达到重富营养化水平。磷含量较高是因为之前这里是磷肥厂。砷矿关闭后，炉子被用来炼硫酸，硫酸用来生产磷肥。叶绿素 a 的含量相反，在区域 6 和区域 7 较高。水体富营养化分级标准如表 3.2 所示。

表 3.2 水体富营养化分级标准

The classification of eutrophication of lakes

级别	营养水平	TP（mg/L）	TN（mg/L）	Chl-a（μg/L）
1	贫营养	0.01	0.20	0.001
2	中营养	0.025	0.50	0.004
3	中富营养	0.05	1.00	0.010
4	富营养	0.10	1.50	0.030
5	重富营养	0.20	2.00	0.065

3.3.1.2 水体中砷含量分布特征

表层水体中的砷含量如图 3.1 所示。从图中可以看出，原水中的砷含量基本高于过滤水中的砷含量。在区域 1，过滤水中的砷含量为 0.62~2.35 μg/L。该区域所有水样中的砷含量均在我国生活饮用水砷标准 10 μg/L 以下，说明该地区上游水体中基本无砷污染。在区域 2，砷含量增加至 19.80 μg/L。在区域 3、区域 4 和区域 5，水体中的砷含量急剧升高，最高可达 3293 μg/L。除距离中心区 0.1 km 处，砷含量为 59.14 μg/L 外，水样中的砷含量基本大于 1791 μg/L，超过我国饮用水砷标准的 5~329 倍，说明这些区域的水体已经受到严重的砷污染。在区域 6，砷含量明显下降，降低至 27.87 μg/L。进入区域 7 之后，砷含量为 10.44~36.10 μg/L（过滤水），略高于我国饮用水标准（10 μg/L）。由此可知，砷矿区中心区下游和尾矿附近，水体中的砷含量较高，且随着距离的加大，砷含量显著下降。因此，可以推断，当地砷矿的开采和冶炼是水体砷污染的重要来源。

进一步分析，不同区域过滤水中的砷含量分别为，区域 1：1.34±0.51 μg/L，区域 2：10.82±7.36 μg/L，区域 3：1707±1576 μg/L，区域 4：2536±216 μg/L，区域 5：2012±185 μg/L，区域 6：126.72±77.49 μg/L，区域 7：20.97±14.45 μg/L。采用 SPSS 对不同区域水体中的砷含量进行多组样本间差异显著性分析，结果如图 3.2 所示，可见区域 3、区域 4、区域 5 水体中的砷含量显著高于区域 1、区域 2、区域 6、区域 7。综上所述，水体中的砷含量随着与中心区和尾矿区距离的加大而呈现明显的降低趋势，说明砷矿的开采等活动是当地砷污染的一个重要来源。

世界各地淡水系统水体中的砷含量如表 3.3 所示。Alvarez 等人（2006）研究发现，西班牙 León 水库上游水体中的砷含量为 33 μg/L，而经过 Santa Águeda 砷矿的下游水体中的砷含量则高达 890 μg/L。Watanabe 等人（2008）研究发现，日本 Ginzan 溪上游水体中的砷含量为 80 μg/L，而经过 Karuizama 矿的下游水体中的砷含量高达 1040 μg/L。Kim 等人（2009）研究发现，韩国临近 Ulsan 铁钨矿的 Cheongog Spring 水体中的砷含量为 277 μg/L，而随着与矿区距离的加大，砷含量显著下降。Culioli 等人（2009）研究发现，法国 Bravona 和 Presa 河上游水体中的砷含量为 2.13 μg/L，而经过砷矿的下游水体中的砷含量高达 2330 μg/L。结果基本与本研究一致，因此，矿采活动造成的环境砷污染需引起重视。

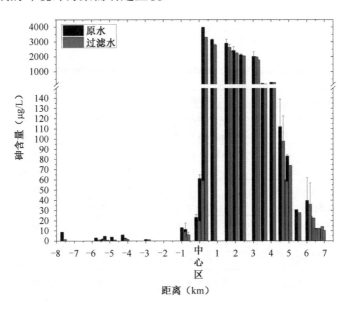

图 3.1　表层水体中的砷含量

The arsenic concentration in surface water

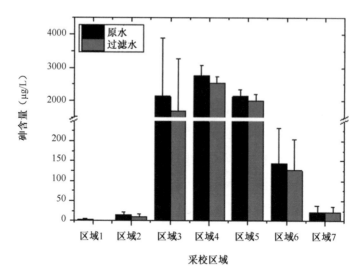

图 3.2　不同区域水体中的砷含量

The arsenic concentration in surface water from different areas

表 3.3　世界各地淡水系统水体中的砷含量

Concentration of arsenic in some freshwater systems in literature

研究区域	砷来源	砷含量（μg/L）	来源
León 水库，西班牙	Santa Águeda 砷矿	33～890	Alvarez et al., 2006
Ginzan 溪，日本	Karuizama 铅锌矿	80～1040	Watanabe et al., 2008
Meadow 溪，美国	Stibnite 砷矿	2～27373	Dovick et al., 2016
Orib 盆地，法国	Bournac 锑矿	19.7～78.3	Casiot et al., 2007
Bravona 和 Presa 河，法国	砷矿	2.13～2330	Culioli et al., 2009
Hayakawa 河，日本	Mt. Hakone 温泉区	17～750	Miyashita et al., 2009
Francolí 河，西班牙	城市工业废水	0.49～3.41	Roig et al., 2013
锡矿山，中国	锡矿山锑矿	0.56～11.3	Fu et al., 2010
Hayle 等河，英国	矿产开采	5～75	Awrahman et al., 2015
Moira 河，加拿大	Deloro 砷铜矿	2～140	Zheng et al., 2003
Yellowknife 湖，加拿大	Con Mine 金矿	7～250	Caumette et al., 2011
太湖和滇池，中国	生产生活	1.36～10.90	Zhang et al., 2013

续表

研究区域	砷来源	砷含量（μg/L）	来源
Kiba 等湖，日本	生产生活	2～48	Hasegawa et al.，2010
Manchar 湖，巴基斯坦	工农业废水	60.4～101.8	Arain et al.，2008
Colongulac 湖，澳大利亚	工农业废水	30～164	Madigan et al.，2005
阳宗海，中国	工业废水	123.8～234.8	Wang et al.，2010
Mohawk 湖，美国	工农业废水	22.1～40.0	Barrringer et al.，2011
Cheongog Spring，韩国	Ulsan 铁钨矿	4～277	Kim et al.，2009

表层水体（过滤水）中的砷形态如图 3.3 所示。表层水体中的砷主要为无机砷——As(V)和 As(Ⅲ)。大部分表层水体中的砷形态只有 As(V)，而在少部分水样中，As(Ⅲ)的比例可达到 16%。这与大多数的研究结果一致。Miyashita 等人（2009）对日本 Hayakawa 河水体中的砷形态的分析中仅检测到 As(V)。同样，Zheng 等人（2003）研究发现，As(V)是加拿大 Moria 河表层水体中主要的砷形态（占比 98%），而 As(Ⅲ)仅占 2%。Dovick 等人（2016）研究发现，As(V)是美国 Stibnite 矿表层水体中主要的砷形态（占比大于 63%）。Ronkart 等人（2007）对天然矿泉水中的砷形态的分析中，大多仅检测到 As(V)，部分水样中检测到微量的 As(Ⅲ)。Casiot 等人（2007）研究发现，Bournac Greek 表层水体中的砷形态主要为 As(V)，占比大于 91%。但是也有大量研究发现，淡水水体中除无机砷 As(V)和 As(Ⅲ)外，还含有一定量的有机砷。Kim 等人（2009）研究发现，韩国 Cheongog Spring 水体中的砷形态主要为 As(V)，部分水样中检测到 As(Ⅲ)、DMA、MMA。Barringer 等人（2011）研究也发现，美国 Mohawk 湖的水体中的砷形态主要为 As(V)、As(Ⅲ)，还含有少量的 MMA、DMA。而 Hasegawa 等人（2010）研究则发现，日本 18 个湖的水体中的砷形态除 As(V)、As(Ⅲ)、MMA、DMA 外，还含有 UV-As、UV-MMA、UV-DMA 等多种复杂的砷形态。值得一提的是，水体中的砷形态受到多种因素的影响，如 pH 值、水体富营养化程度、季节、水温等（Hasegawa et al.，2010；杨芬 等，2015）。

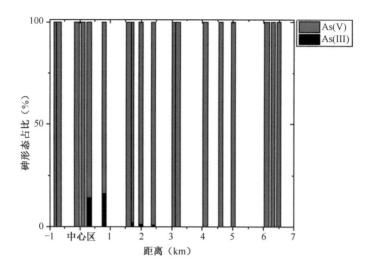

图 3.3　表层水体中的砷形态

The arsenic speciation in surface water

3.3.2　沉积物中砷的分布特征

3.3.2.1　沉积物中理化性质分布特征

沉积物的基本理化性质如表 3.4 所示，不同区域的理化性质变化范围较小。沉积物基本呈弱碱性。沉积物的总有机质在区域 3 最低，仅为 6 g/kg 左右，区域 5、区域 6、区域 7 的总有机质含量较高。

表 3.4　沉积物的基本理化性质

The basic property of soils in each sampling area

区域	总有机质（g/kg）	pH 值
2	13.67±0.47	7.73±0.14
3	6.16±0.32	7.83±0.08
4	16.91±7.04	7.67±0.27
5	19.85±1.89	7.51±0.26
6	21.82±4.93	7.51±0.14
7	21.88±5.51	7.30±0.34

3.3.2.2　沉积物中砷的分布特征

沉积物中的砷含量如图 3.4 所示，砷含量范围为 9.53～4543 mg/kg。在中心区 1 km 内，沉积物中的砷含量为 109～410 mg/kg。而在尾矿坝周围（1.5～2.0 km），沉积物中的砷含量最高，范围为 1728～4543 mg/kg。在尾矿坝下游和黄水溪之间（2.0～5.5 km），沉积物中的砷含量为 98～612 mg/kg。在区域 7（大于 5.5 km），沉积物中的砷含量为 9～137 mg/kg。根据我国土壤环境质量标准，除区域 7 的极少土壤中砷含量小于 35 mg/kg 外，绝大多数土壤中砷含量严重超标。

这与很多砷污染区中土壤的砷含量基本一致。Kim 等人（2009）报道，随着距离 Ulsan 矿的距离加大，沉积物中的砷含量从 260 mg/kg 下降至 116 mg/kg。Culioli 等人（2009）研究发现，受旧砷矿采矿活动影响，法国 Bracona 和 Presa 河沉积物中的砷含量为 216～9135 mg/kg。Barringer 等人（2011）研究则发现，新西兰 Mohawk 湖表层沉积物中的砷含量为 91～460 mg/kg。Zheng 等人（2003）研究发现，加拿大 Moria 河底沉积物中的砷含量为 138～528 mg/kg。Zhang 等人（2013）研究发现，我国太湖和滇池沉积物中的砷含量为 4.66～169.25 mg/kg。Wang 等人（2010）研究发现，阳宗海沉积物中的砷含量为 11.47～164.70 mg/kg。Dovick 等人（2016）研究发现，美国

Stibnite 矿沉积物中的砷含量从 4.4~19.9 mg/kg 增加至 149~15192 mg/kg。

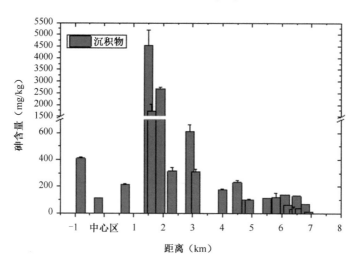

图 3.4　沉积物中的砷含量

The arsenic concentration in sediment

沉积物是可能的砷贮存库，研究沉积物中间隙水中的砷有助于阐明水—沉积物界面砷的迁移过程。在本研究中，间隙水中的砷含量如图 3.5 所示。在中心区上游 1 km（区域 2）内，间隙水中的砷含量为 21.12~24.59 μg/L。而在下游和尾矿坝周围（0.5~2.0 km），间隙水中的砷含量达到 1755~6472 μg/L。在区域 6 和区域 7（大于 3.5 km），间隙水中的砷含量急剧下降，为 2.60~28.37 μg/L。综上所述，间隙水中的砷含量随着与中心区距离的加大而呈现明显的降低趋势。这也说明，砷矿开采活动是当地砷污染的主要原因。

间隙水中的砷形态分析结果如图 3.6 所示。间隙水中的砷主要为无机砷 As(Ⅴ)和 As(Ⅲ)，部分间隙水中仅含有 As(Ⅲ)。间隙水处在一种缺氧状态，因此相比表层水体，As(Ⅲ)的比例明显增加。Zheng 等人（2003）研究发现，As(Ⅲ)是加拿大 Moria 河底层水体中主要的砷形态（占比 70%）。

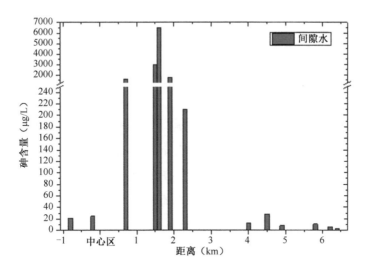

图 3.5 间隙水中的砷含量

The arsenic concentration in porewater

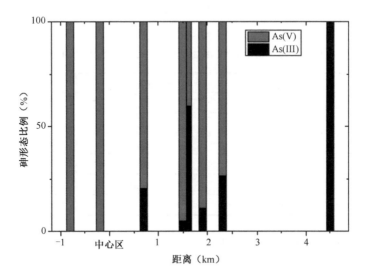

图 3.6 间隙水中的砷形态分析结果

The arsenic speciation in porewater

3.3.3 土壤中砷的分布特征

3.3.3.1 土壤的理化性质的分布特征

土壤的基本理化性质如表 3.5 所示,不同区域的理化性质差异明显。区域 1 和区域 2 土壤呈弱酸性,pH 值分别为 6.25 和 6.30,而区域 3、区域 4、区域 5、区域 6 的土壤呈弱碱性 pH 值范围为 7.08~7.62。同样,土壤含水率从 15.72%到 34.67%。土壤总有机质变化范围也比较广,区域 2 和区域 6 总有机质含量较低,分别为 16.84 g/kg 和 18.87 g/kg,区域 3 和区域 4 含量较高,分别为 33.41 g/kg 和 29.02 g/kg。这主要与采集的土壤类型有关,区域 3 和区域 4 以农田为主,区域 2 和区域 6 以尾矿土和河床土为主。

表 3.5 土壤的基本理化性质(平均值)

The basic property of soils in each sampling area

	含水率(%)	总有机质(g/kg)	pH 值
区域 1	—	22.76	6.25
区域 2	15.72	16.84	6.30
区域 3	23.34	33.41	7.32
区域 4	18.07	29.02	7.08
区域 5	27.68	24.29	7.26
区域 6	34.67	18.87	7.62

3.3.3.2 土壤中砷的分布特征

本研究中,采集的土壤中的砷含量范围为 7.32~5008 mg/kg,远远高于湖南省土壤砷背景值(13.41 mg/kg)。如图 3.7 所示,中心区和尾矿区的土壤中的砷含量最高,随着与中心区和尾矿区的距离增加,土壤中的砷含量逐渐降低。在区域 1 和区域 2,土壤中的砷含量为 7.32~212 mg/kg。在

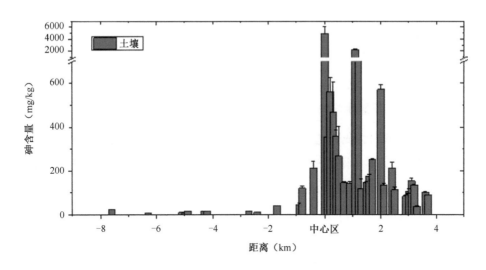

图 3.7 土壤中的砷含量

The arsenic concentration in soil

区域 3，土壤中的砷含量为 142～5008 mg/kg。在区域 4，土壤中的砷含量为 116～2224 mg/kg。曾敏等人（2006）也发现，湖南石门尾砂覆盖土中的砷含量可达到 7097 mg/kg。在区域 5 和区域 6，土壤中的砷含量为 35～211 mg/kg。以上说明砷矿的开采等活动是当地土壤砷污染的重要来源。同样，关于矿区土壤砷污染的报道很多。Alvarez 等人（2006）研究发现，西班牙 Santa Águeda 砷矿土壤中的砷含量可达 23800 mg/kg。Thouin 等人（2016）研究发现，Place-à-Gaz 地区土壤中的砷含量异常高，范围为 1937～72820 mg/kg。Milton 和 Johnson（1999）研究发现，爱尔兰一尾矿池的平均砷含量为 630±34 mg/kg。本研究中，土壤中的砷含量也基本高于湖南省其他地区土壤中的砷含量。Ma 等人（2017）和 Liang 等人（2017）对湖南省土壤采集分析，得出砷含量基本小于 25 mg/kg。Ding 等人（2017）采集湖南省郴州市土壤，其砷含量为 1.31～443 mg/kg。Fu 等人（2016）研究发展湖南省锡矿山尾矿中的砷含量最高可达 853 mg/kg。根据我国土壤环境质量标准，III类土壤中砷含量应小于等于 40 mg/kg，本研究中采集的土壤，其中约 99%其砷含量严重超标。长期重金属污染会导致土壤养分流失，破坏土壤生物，导致土壤功能退化（Zhang et al.，2016；Zhao et al.，2013）。因此，石门雄黄矿土壤砷污染治理势在必行。

2016 年 6 月，我们选取了 4 个点位的表层土壤剖面，测定砷含量，结果如图 3.8 所示。剖面分为 8 层：0～15 cm 为 a 层；15～30 cm 为 b 层；30～45 cm 为 c 层；45～60 cm 为 d 层；60～75 cm 为 e 层；75～90 cm 为 f 层；90～105 cm 为 g 层；105～120 cm 为 h 层。砷矿区上游 6.3 km 的土壤中的砷含量为 6.97～9.40 mg/kg。而下游 0.1 km 的土壤中的砷含量则出现随着深度的增加而不断增加的趋势，范围为 273～658 mg/kg。下游 2.2 km 和 1.2 km 的土壤中的砷含量则随着深度的增加而不断增加，但是在 h 层，砷含量显著降低。该研究结果与研究报道中的基本一致。Tang 等人（2016）研究发现，石门雄黄矿尾矿区和矿区土壤中的砷含量会随着深度的增加先增加后下降。Li 等人（2017）分析中国香港地区的土壤剖面中的砷也发现，土壤中的砷含量基本上先增加后降低。Perkins 等人（2016）研究发现，爱尔兰尾矿修复土壤剖面中土壤表层的砷含量远远低于深层砷含量。Falinski 等人（2014）研究发现，夏威夷 Kaumana 保护区土壤中的砷含量在 0～30 cm 深度含量高，但是在大于 30 cm 的深度，砷含量会急剧下降。在本研究中，有可能是暴雨天气加大了砷的下渗，从而出现随着土壤深度的增加砷含量不断增加的现象。一般而言，土壤中砷的迁移速度十分缓慢，从而导致土壤污染长期难以减轻。该现象应在土壤砷修复过程中引起重视。

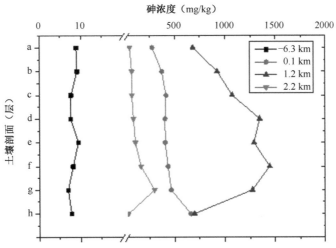

图 3.8 土壤剖面中的砷含量

The arsenic concentration in soil profiles

3.4 本章小结

本章对石门雄黄矿及周边地区水体、沉积物、土壤中的砷污染程度及分布特征进行了研究。水体、沉积物和土壤中的砷含量最高分别可达到 3293 μg/L、4543 mg/kg、5008 mg/kg。结果表明，环境中的砷含量随着与矿区和尾矿区距离的加大而明显降低，说明受原砷矿采矿和冶炼活动的影响，当地环境受到了严重的砷污染。

参考文献

黄益宗, 郝晓伟, 雷鸣, 等. 重金属污染土壤修复技术及其修复实践[J]. 农业环境科学学报, 2013, 32:409-417.

王振刚, 何海燕, 严于伦. 石门雄黄矿地区居民砷暴露研究[J]. 卫生研究, 1999, 28:6-8.

杨芬, 朱晓东, 韦朝阳. 陆地水环境中砷的迁移转化[J]. 生态学杂志, 2015, 34:1448-1455.

曾敏, 廖柏寒, 曾清如. 湖南郴州、石门、冷水江 3 个矿区 As 污染状况的初步调查[J]. 农业环境科学学报, 2006, 25:418-421.

ALVAREZ A, ORDÓÑEZ A, LOREDO J. Geochemical assessment of an arsenic mine adjacent to a water reservoir (León, Spain)[J]. Environ Geol, 2006, 50:873-884.

ARAIN M B, KAZI T G, JAMALI M K, et al. Total dissolved and bioavailable elements in water and sediment samples and their accumulation in *Oreochromis mossambicus* of polluted Manchar Lake[J]. Chemosphere, 2008, 70:1845-1856.

AWRAHMAN Z A, RAINBOW P S, SMITH B D, et al. Bioaccumulation of arsenic and silver by the caddisfly larvae *Hydropsyche siltalai* and *H. pellucidula*: A biodynamic modeling approach[J]. Aquat Toxicol, 2015, 161:196-207.

BARRINGER J L, SZABO Z, WILSON T P, et al. Distribution and seasonal dynamics of arsenic in a shallow lake in northwestern New Jersey, USA[J]. Environ Geochem Health, 2011, 33:1-22.

BRINTON W F. An international look at compost standards[J]. Biocycle,

2001, 42:74-76.

CASIOT C, UJEVIC M, MUNOZ M, et al. Antimony and arsenic mobility in a creek draining an antimony mine abandoned 85 years ago (upper Orb basin, France)[J]. Appl Geochem, 2007, 22:788-798.

CAUMETTE G, KOCH I, ESTRADA E, et al. Arsenic speciation in plankton organisms from contaminated lakes: transformations at the base of the freshwater food chain[J]. Environ Sci Technol, 2011, 45:9917-9923.

CULIOLI J L, FOUQUOIRE A, CALENDINI S, et al. Trophic transfer of arsenic and antimony in a freshwater ecosystem: A field study[J]. Aquat Toxicol, 2009, 94:286-293.

DOVICK M A, KULP T R, ARKLE R S, et al. Bioaccumulation trends of arsenic and antimony in a freshwater ecosystem affected by mine drainage[J]. Environ Chem, 2016, 13:149-159.

FALINSKI K A, YOST R S, SAMPAGA E, et al. Arsenic accumulation by edible aquatic macrophytes [J]. Ecotox Environ Safe, 2014, 99:74-81.

FU Z, WU F, AMARASIRIWARDENA D, et al. Antimony, arsenic and mercury in the aquatic environment and fish in a large antimony mining area in Hunan, China[J]. Sci Total Environ, 2010, 408:3403-3410.

FU Z, WU F, MO C, et al. Comparison of arsenic and antimony biogeochemical behavior in water, soil and tailings from Xikuangshan, China[J]. Sci Total Environ, 2016, 539:97-104.

HASEGAWA H, AZIZUR RAHMAN M, KITAHARA K, et al. Seasonal changes of arsenic speciation in lake waters in relation to eutrophication[J]. Sci Total Environ, 2010, 408:1684-1690.

HUGHES M F, BECK B D, CHEN Y, et al. Arsenic exposure and

toxicology: A historical perspective[J]. Toxicol Sci, 2011, 123:305-332.

KIM Y T, YOON H O, YOON C, et al. Arsenic species in ecosystems affected by arsenic-rich spring water near an abandoned mine in Korea[J]. Environ Pollut, 2009, 157:3495-3501.

LIANG J, FENG C, ZENG G, et al. Spatial distribution and source identification of heavy metals in surface soils in a typical coal mine city, Lianyuan, China[J]. Environ Pollut, 2017, 225:681-690.

LI J S, BEIYUAN J, TSANG D C W, et al. Arsenic-containing soil form geogenic source in Hong Kong: Leaching characteristics and stabilization/solidification[J]. Chemosphere, 2017, 182:31-39.

MA L, WANG L, JIA Y, et al. Accumulation, translocation and conversion of six arsenic species in rice plants grown near a mine impacted city[J]. Chemosphere, 2017, 183:44-52.

MADIGAN B A, TUROCZY N, STAGNITTI F. Speciation of arsenic in a large endoheric lake[J]. Bull Environ Contam Toxico, 2005, 75:1107-1114.

MANDAL B K, SUZUKI K T. Arsenic round the world: a review[J]. Talanta, 2002, 58:201-235.

MAUNOURY-DANGER F, FELTEN V, BOJIC C, et al. Metal release from contaminated leaf litter and leachate toxicity for the freshwater crustacean *Gammarus fossarum*[J]. Environ Sci Pollut R, 2017, doi:10.1007//s11356-017-9452-0.

MDEGELA R H, BRAATHEN M, PEREKA A E, et al. Heavy metal and organochlorine residues in water, sediment, and fish in aquatic ecosystems in urban and peri-urban areas in Tanzania[J]. Water Air Soil Pollut, 2009, 203:369-379.

MILTON A, JOHNSON M. Arsenic in the food chains of a revegetated metalliferous mine tailing pond[J]. Chemosphere, 1999, 39:765-779.

MIYASHITA S, SHIMOYA M, KAMIDATE Y, et al. Rapid determination of arsenic species in freshwater organisms from the arsenic-rich Hayakawa River in Japan using HPLC-ICP-MS[J]. Chemosphere, 2009, 75:1065-1073.

PERKINS W T, BIRD G, JACOBS S R, et al. Field-scale study of the influence of differing remediation strategies on trace metal geochemistry in metal mine tailings from the Irish Midlands[J]. Environ Sci Pollut Res, 2016, 23:5592-5608.

ROIG N, SIERRA J, ORTIZ J D, et al. Integrated study of metal behavior in Mediterranean stream ecosystems: A case-study[J]. J Hazard Mater, 2013, 263P:122-130.

RONKART S N, LAURENT V, CARBONNELLE P, et al. Speciation of five arsenic species (arsenite, arsenate, $MMAA^V$, $DMAA^V$ and AsBet) in different kind of water by HPLC-ICP-MS[J]. Chemosphere, 2007, 66:738-745.

TANG J W, LIAO Y P, YANG Z H, et al. Characterization of arsenic serious-contaminated soils from Shimen realgar mine area, the Asian largest realgar deposit in China[J]. J Soil Sediment, 2016, 16,1519-1528.

THOUIN H, LE FORESTIER L, GAUTRET P, et al. Characterization and mobility of arsenic and heavy metals in soils polluted by the destruction of arsenic-containing shells from the Great War[J]. Sci Total Environ, 2016, 550:658-669.

WANG F X, CHAPMAN P M. Biological implications of sulfide in sediment: A review focusing on sediment toxicity[J]. Environ Toxicol Chem, 1999, 18:2526-2532.

WANG Z H, HE B, PAN X J, et al. Levels, trends and risk assessment of arsenic pollution in Yangzonghai Lake, Yunnan Province, China[J]. Sci China Chem, 2010, 53:1809-1817.

WATANABE K, MONAGHAN M T, TAKEMON Y, et al. Biodilution of heavy metals in a stream macroinvertebrate food web: Evidence from stable isotope analysis[J]. Sci Total Environ, 2008, 394:57-67.

ZHANG N, WEI C, YANG L. Occurrence of arsenic in two large shallow freshwater lakes in China and a comparison to other lakes around the world[J]. Microchem J, 2013,110:169-177.

ZHANG C, NIE S, ZENG G, et al. Effects of heavy metals and soil physicochemical properties on wetland soil microbial biomass and bacterial community structure[J]. Sci Total Environ, 2016, 557-558:785-790.

ZHAO L, XU Y, HOU H, et al. Source identification and health risk assessment of metals in urban soils around the Tanggu chemical industrial district, Tianjin, China[J]. Sci Total Environ, 2013, 468-469C:654-662.

ZHENG J, HINTELMANN, DIMOCK B, et al. Speciation of arsenic in water, sediment, and plants of the Moira watershed, Canada, using HPLC coupled to high resolution ICP-MS[J]. Anal Bioanal Chem, 2003, 377:14-24.

ZHU X, WANG R, LU X, et al. 2015. Secondary minerals of weathered orpiment-realgar-bearing tailings in Shimen carbonate-type realgar mine, Changde, Central China. Miner Petrol, 2015,109:1-15.

第 4 章

陆地环境中砷的生物富集与转化特征

第4章 陆地环境中砷的生物富集与转化特征

4.1 陆地环境中砷的生物富集与转化研究意义

砷是环境中普遍存在的毒性最强的元素之一（Mandal and Suzuki, 2002; Ng, 2005）。砷的毒性不仅与总量有关，还取决于化学形态特征（Hughes, 2002; Shen et al., 2016）。因此，对环境和生物体中砷形态的研究十分重要。采矿和冶炼活动是土壤砷污染的一个重要原因（Smedley and Kinniburgh, 2002; Garelick et al., 2008）。土壤中的砷会影响生态系统的功能和结构，砷污染已得到广泛关注（Maunoury-Danger et al., 2017）。植物可以吸收土壤中的砷，甚至可以在体内富集（Ma et al., 2001; Wei et al., 2015; Wan et al., 2017）。陆地植物中的砷形态主要为 As(Ⅴ)、As(Ⅲ)、DMA 和 MMA（Amaral et al., 2013）。但是关于全面报道砷矿区本土植物中的砷特征的研究相对少见（Otones et al., 2011）。

凋落物分解是生态系统中养分和能量循环的重要过程（Jonczak et al., 2014; Berg and McClaugherty, 2014）。在植物的生长过程中，通过凋落物分解释放的砷不可忽视（Maunoury-Danger et al., 2017）。蚯蚓占土壤动物生物量的 80%，且常年生活在土壤中（Sivakumar, 2015），在土壤系统中起着重要作用（van Gestel et al.，2009）。有研究发现，蚯蚓能够提高土壤中水溶性砷含量，从而提高植物对砷的耐受性（Sizmur et al.，2011a；Jusselme et al.，2013）。蚯蚓可通过消化道和皮肤吸收并富集土壤和凋落物中的砷，且方便采集，被视为土壤污染常用的生物标志物（Shin et al.，2007；Dada et al.，2013）。因此，蚯蚓作为土壤环境的指示生物的研究越来越广泛（Sizmur et al.，2011b；Calisi et al.，2013）。但是，大多数研究主要基于室内模拟研究，野外研究相对少见（Calisi et al.，2013；Wang et al.，2016）。

蚯蚓和植物位于陆地生态食物链的底端，是高级生物的潜在食物来源，如鸟类、鸡等（Moriarty et al.，2009；Button et al.，2011；2012）。值得一提的是，砷暴露会影响鸟类的繁殖，导致生育力降低、死亡率提高（Janssens et al.，2003；Belskii et al.，2005；Eeva et al.，2009）。其中，雀形目鸟由于分布广泛，对环境变化敏感，常常作为环境砷污染的生物标志物（Sánchez-Virosta et al.，2015）。但是关于雀形目鸟中砷特征的报道十分少见（Koch et al.，2005；Sánchez-Virosta et al.，2015）。关于野生鸟类中砷的生物转化的研究极其有限（Moriarty et al.，2009）。

鸡是被消费最多的肉类之一，鸡蛋、鸡心、鸡胗、鸡肝等深受人们的喜爱。另外，鸡胗皮（鸡内金）在我国传统医学中有治疗饮食积滞、肾虚遗精、夜间盗汗等疾病的功能。研究表明，2014年中国大陆和亚洲养殖的鸡分别占世界的21.3%和55.1%（FAOSTAT，2014）。关于鸡体内砷形态及砷暴露的研究对象主要为农场生产的肉鸡（Pizarro et al.，2015；Liu et al.，2016；Hu et al.，2017）。在养殖业中，洛克沙肿（ROX）由于能提高饲料效率、促进家禽生长、控制肠道寄生虫，常常被用作饲料添加剂（Nachman et al.，2013；Liu et al.，2015a）。尽管农场养殖是商业肉类的主要来源，但是近年来无公害的散养鸡得到越来越受城乡居民的喜欢（Grace and Macfarlane，2016）。到目前为止，关于散养鸡中砷特征的研究仍未见报道。

总的来说，在陆地生态系统中，土壤中的砷能通过根系进入植物体中，通过凋落物分解释放回土壤中，土壤动物吸收土壤和凋落物中的砷（Pulford and Watson，2003；Dedeke et al.，2016；Wang et al.，2016）。植物和土壤动物又是食物链顶端高营养级生物如鸟类和鸡的潜在食物来源。然而，大量的文献仅仅关注土壤—植物，土壤—蚯蚓系统中砷的迁移（Van Nevel et al.，2014；da Silva Souza et al.，2014；Lucisine et al.，2015；Kramar et al.，2017）。本研究对土壤—植物—凋落物—蚯蚓—高营养级生物系统中的砷富集和转化的研究具有重要意义。本研究的目的是：厘清砷矿区不同介质（土壤、植物、凋落物、蚯蚓、鸟类、鸡）中的总砷和砷形态；通过分析食物来源，对陆地生态系统中某些有机砷的潜在来源和途径进行分析。

4.2　砷矿区陆地环境生物样品采集、总砷及砷形态分析方法

2015 年 5 月和 2016 年 6 月，我们采集了土壤、蚯蚓、蜗牛、凋落物、植物等样品。共 17 个采样点，其中 6 个采样点（L～Q）位于区域 1，11 个采样点（A～K）位于区域 3、区域 4 和区域 5。采样点详细信息如表 4.1 所示。本研究中，草本植物为混合物，主要有狼尾草（*Pennisetum alopecuroides*）、马唐草（*Digitaria sanguinalis*）、狗牙根（*Cynodon dactylon*）、牛筋草（*Eleusine indica*）和车前草（*Plantago asiatica*）。样品的采集、预处理和实验分析见第 2 章。

表 4.1 采样点详细信息

Site description 和 physico-chemical properties of soils

		采样点	距离[①]	描述	砷含量 (mg/kg) (平均值)	pH 值 (平均值)	TOM (g/kg) (平均值)
污染区	区域 5	A	+3.1	草地	152.00	7.95	25.99
		B	+3.0	草地	92.18	6.57	24.23
		C	+2.9	草地	83.29	6.65	21.43
		D	+2.5	农家后院	110.00	7.51	18.05
		E	+2.2	农业用地	61.74	5.93	30.81
	区域 4	F	+1.7	山丘	12.76	5.78	41.36
		G	+1.2	山丘	893.00	4.51	40.04
		H	+1.1	废弃地	2224.00	7.98	29.51
	区域 3	I	+0.7	农家后院	144.00	6.80	49.33
		J	+0.5	农业用地	169.00	6.34	19.16
		K	+0.1	农业用地	352.00	5.88	21.07
非污染区	区域 1	L	-4.2	农业用地	15.07	6.45	26.00
		M	-4.3	农业用地	15.48	6.27	22.11
		N	-4.9	农业用地	15.10	6.21	16.12
		O	-5	山丘	8.93	6.45	19.89
		P	-6.3	山丘	8.55	6.74	20.39
		Q	-7.6	山丘	23.75	6.63	24.43

注：[①]采样点与原砷矿中心的距离（单位：km），+代表下游，-代表下游。

我们共获得 6 只鸟样本，包括画眉、白头翁、麻雀和喜鹊（见表 4.2）。所有鸟均为成鸟，且均为雀形目鸟，寿命少于 10 年；均为留鸟，具有固定的生活范围；均为杂食性，主要食物为昆虫和植物种子。解剖这 6 只鸟，获取肌肉和羽毛，肌肉为腿部肌肉。对于喜鹊，由于其体型比较大，除获取肌肉和羽毛外，还需获取肝脏、心脏、胃、胃内容物和爪。

表 4.2 鸟样本的相关信息

Details of birds collected in this study

采样时间	鸟类	拉丁文名	体长（cm）	体重（g）	$\delta^{13}C$（‰）	$\delta^{15}N$（‰）
2015 年	画眉	Leucodioptron canorus	15	33	−24.42±0.12	8.01±0.09
	麻雀 1	Passer montanus	16	30	−25.51±0.01	8.29±0.03
	麻雀 2	Passer montanus	15	10	−24.60±0.09	5.53±0.02
	白头翁 1	Pycnonotus sinensis	11	17	−26.04±0.06	9.57±0.17
2016 年	白头翁 2	Pycnonotus sinensis	10	40	−25.49±0.06	9.18±0.08
	喜鹊	Pica pica	20	159	−26.13±0.10	8.12±0.02

2016 年 6 月，我们分别从两个农家后院（GY 和 CY）采集了 4 只家养母鸡（见表 4.3）。采集地点位于石门雄黄矿上游约 100 m 处，但是两处的地形相差较大：GY 靠近小山丘，而 CY 在黄水溪边。所有母鸡均为散养模式，农家后院面积约 50 m²。母鸡的食物主要为植物、昆虫，吃进去的还有砂砾，饮水为自来水，CY 处的饮水来自黄水溪。每只母鸡解剖出以下组织/器官：肌肉、羽毛、肝脏、心脏、鸡胗、鸡胗皮、胃内容物、鸡卵。

表 4.3 家禽母鸡的相关信息

Details of chickens collected from the two backyards

点位	家养母鸡					砷含量		
	编号	体重（kg）	年龄	$\delta^{13}C$（‰）	$\delta^{15}N$（‰）	玉米[①]	大米[①]	自来水[②]
GY	GY1	2.50	3	−15.16±0.14	6.60±0.17	0.02	0.07	3.50
	GY2	2.00	3	−13.98±0.23	6.92±0.04			
CY	CY1	2.25	3	−19.52±0.11	7.78±0.03	0.15	0.18	3.66
	CY2	1.75	3	−20.67±0.09	6.65±0.08			

注：[①]单位为 mg/kg；[②]单位为 μg/L。

所有组织/器官用自来水清洗多次，再用去离子水淋洗，用自封袋装，编号后冷冻保存，运回实验室。之后在-80℃下冷冻干燥。肌肉组织用来测定碳氮同位素。其他样品均在液氮下通过 Tissuelyser-24 全自动样品快速研磨仪（上海净信科技有限公司生产）研磨均匀，在-10℃下保存。样品的实验分析方法见第 2 章。

4.3 陆地环境中砷的生物富集与转化特征

4.3.1 土壤—植物—凋落物—土壤动物系统中砷的富集特征

4.3.1.1 土壤中砷的分布特征

如表 4.1 所示,在污染区,pH 值变化范围从 4.51 到 7.98,TOM 变化范围从 18.05 g/kg 到 49.33 g/kg。而在非污染区,pH 值变化范围较小(6.21~6.74),而 TOM 变化范围从 16.12 g/kg 到 26.00 g/kg。在非污染区,砷含量为 8.55~23.75 mg/kg,远远低于污染区的砷含量(61.74~2224.00 mg/kg,除采样点 F 外)。采样点 F 位于山丘边上,可能受到了沉积作用,砷含量与非污染区区别不大。值得注意的是,砷矿区和尾矿坝附近的采样点,包括 G、H、I、J 和 K,其土壤中的砷含量高于我国土壤环境质量标准中Ⅲ类土壤中的砷含量标准限值(40 mg/kg)的 3.6~55.6 倍。结果表明,多年的砷矿开采和冶炼是导致当地砷严重污染的重要原因。

4.3.1.2 植物中砷的分布特征

在不同采样点，不同种类陆地植物的不同组织（根、叶、茎和其他）中的砷含量及生物富集系数（BF）差异较大（见图4.1）。总的来说，砷含量表现为叶大于茎，该结果与其他研究一致（Rosas et al., 2014; Tsipoura et al., 2017）。Otones 等人（2011）研究发现，西班牙 Barruecopardo 矿区土壤中的砷含量为 70~5330 mg/kg，其陆地植物的 BF 为 0.00005~0.054。Freitas 等人（2004）研究发现，葡萄牙 São Domingos 铜矿区（土壤砷含量为 72~607 mg/kg）的陆地植物的 BF 为 0.0006~0.032。Larios 等人（2012）研究发现，西班牙两个汞矿区（土壤砷含量为 1260~25901 mg/kg）的植物的 BF 为 0.01~0.1。

在污染区采集的蜈蚣草和大叶井口边草中的砷含量最高，它们的地上部分砷含量分别达到 584 mg/kg 和 391 mg/kg。BF 可用来评估植物将土壤中的砷转移到植物体内的能力，对于超富集植物而言，BF 一般大于 1（Brooks，1998）。在本研究中，蜈蚣草的 BF 为 0.65~2.16，而大叶井口边草的 BF 为 1.11，说明两者均对砷具有超富集能力。该结果与污染区土壤中两种植物的研究结果一致（Cao et al., 2004; Wei et al., 2006; Jeong et al., 2014）。橘树和大多数蔬菜中的砷含量为 0.005~6.92 mg/kg，除茄子的 BF 达到 0.20 外，其他植物的 BF 变化范围从 0.0003 到 0.039，这些植物并没有超富集砷的能力。

玉米是我国广泛种植的农作物之一，也是主要食用的粮食作物之一。在本研究中，采样点 K 的玉米叶片中的砷含量为 6.9 mg/kg，茎中的砷含量为 3.1 mg/kg，远远高于非污染区（小于 0.5 mg/kg）。同样的，在污染区，芒草和小蓬草中的砷含量（0.52~5.52 mg/kg）远远高于非污染区（0.42~0.57 mg/kg）。因此，尽管石门雄黄矿已经关闭 15 年，但是过去的采矿和冶炼活动造成了本土植物不同程度的砷富集问题。这一发现正好说明陆地生态系统中砷的富集和转化研究的重要性。

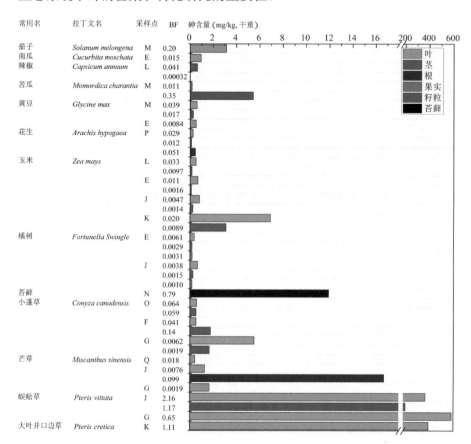

图 4.1 不同组织中的砷含量及生物富集系数

Arsenic concentrations in plant tissues 和 bioaccumulation factor

4.3.1.3 草本植物—凋落物—蚯蚓系统中砷的分布特征

土壤—草本植物—凋落物—蚯蚓系统中的砷含量及蚯蚓对砷的富集系数如图 4.2 所示，其中包括 3 个非污染区采样点和 7 个污染区采样点。采样点

J 采集到的蜗牛体内砷含量为 2.26～5.19 mg/kg，远远低于蚯蚓体内砷含量。在非污染区，土壤中的砷含量为 8～15 mg/kg，蚯蚓体内砷含量为 5.6～7.8 mg/kg。在污染区，土壤中的砷含量为 83～169 mg/kg，蚯蚓体内砷含量为 31～71 mg/kg，并表现为蚯蚓体内砷含量随着土壤中的砷含量的增加呈现先增加后降低的趋势。值得注意的是，采样点 I 土壤中的 TOM 最高（49.33 g/kg），但是其 pH 值（6.70）低于采样点 D（7.51）和采样点 A（7.95）。这说明，土壤性质如 pH 值和 TOM，会影响土壤中的有效砷含量，这与前人研究结果一致（Niazi et al., 2011；Romero-Freire et al., 2015）。当土壤中的砷含量为 2224 mg/kg（采样点 H）时，蚯蚓体内砷含量最高，达到 430 mg/kg。总的来说，蚯蚓体内砷含量随着土壤中的砷含量的增加而增加。

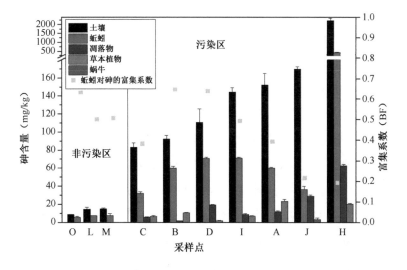

图 4.2　土壤—草本植物—凋落物—蚯蚓系统中的砷含量及蚯蚓对砷的富集系数

Arsenic concentrations in different environmental media 和 bioaccumulation factor of earthworms in soil-grass-litter-invertebrate system

Romero-Freire 等人（2015）发现，当对照组土壤中的砷含量为 3.4～25.7 mg/kg 时，蚯蚓体内砷含量为 32.8 mg/kg。在土壤中分别添加 50 mg/kg、100 mg/kg、300 mg/kg 和 600 mg/kg 砷时，蚯蚓体内砷含量分别达到 319 mg/kg、430 mg/kg、594 mg/kg 和 768 mg/kg。同样，García-Gómez 等人（2014）发现，当对照组土壤中的砷含量为 5±2 mg/kg 时，蚯蚓体内砷含量为 18±7 mg/kg。当土壤中的砷含量从 14.6 mg/kg 增加到 700 mg/kg 时，蚯蚓体内砷含量从 5.5 mg/kg 增加到 5660 mg/kg。Button 等人（2009）也发现，当土壤中的砷含量分别为 289～13080 mg/kg 和 255～3995 mg/kg 时，*Lumbricus rubellus* 和 *Dendrodrilus rubidus* 体内砷含量分别为 11～877 mg/kg 和 15～737 mg/kg。

在污染区，蚯蚓对砷的富集系数（BF）范围为 0.19～0.65。除采样点 C 外，所有采样点的 BF 均呈现随着土壤砷含量增加而降低的趋势。在非污染区，蚯蚓对砷的 BF 为 0.50～0.63，远远高于重污染区（土壤中的砷含量大于 144 mg/kg）。总的来说，BF 随着土壤砷含量的增加而降低，这也说明蚯蚓具有一定自我调节并隔绝砷的能力。纵观前人的研究，蚯蚓对砷的 BF 范围较大。Button 等人（2011）报道，在英国诺丁汉生长的 *Lumbricus terrestris* 对砷的 BF 为 0.37（土壤中的砷含量为 16 mg/kg）。Button 等人（2012）报道，当土壤中的砷含量为 1400 mg/kg 时，*Dendrodrilus rubidus* 对砷的 BF 为 0.31。Watts 等人（2008）报道，当土壤中的砷含量分别为 16～12466 mg/kg 和 16～1005 mg/kg 时，*Lumbricus rubellus* 和 *Dendrodrilus rubidus* 对砷的 BF 分别为 0.04～0.41 和 0.05～0.44。

尽管在这些研究中，蚯蚓对砷的 BF 均小于 1，但是也有研究发现了相对较高的 BF。Lee 等人（2013）报道，当土壤中的砷含量为 18～2297 mg/kg 时，*Eisenia fetida* 对砷的 BF 变化范围从 0.26 到 1.97。同样的，Button 等人（2012）发现，在加拿大 Lower Seal Harbor 土壤（砷含量为 880～2700 mg/kg）中，*Lumbricus castaneus* 对砷的 BF 为 0.49～1.59。Fu 等人

（2011）报道，锡矿山土壤中的砷含量小于 120 mg/kg 时，*Pheretima aspergillum* 对砷的 BF 可达到 6。Romero-Freire 等人（2015）研究发现，当土壤中的砷含量为 3.4～600 mg/kg 时，*Eisenia andrei* 对砷的 BF 为 0.83～11.1。这些研究结果表明，蚯蚓对砷的吸收富集除取决于土壤性质外，也取决于蚯蚓种类。

蚯蚓富集砷的能力远远高于草本植物和凋落物。草本植物和凋落物中的砷含量在同一数量级上，分别为 2.04～23.14 mg/kg 和 1.69～62.42 mg/kg。在某些采样点（如 D、H），凋落物中的砷含量远远高于草本植物，可能是因为凋落物直接吸收了土壤中的砷，或者是叶片在脱落前从幼叶到衰老的过程中累积了砷（Piczak et al., 2003）。相关性分析结果表明，土壤和蚯蚓（R^2=0.989，p=0.001），蚯蚓和凋落物（R^2=0.888，p=0.008），土壤和凋落物（R^2=0.913，p=0.004），两两之间显著相关，说明土壤—凋落物—蚯蚓之间的砷迁移存在密切关联。

4.3.2 土壤—植物—凋落物—土壤动物系统中砷的形态特征

4.3.2.1 土壤中砷形态的分布特征

在本研究中，污染区和非污染区的土壤中 XANES 归一化处理谱均出现明显的 As(Ⅴ)峰见图 4.3。图中实线代表原谱图，虚线代表线性组分拟合后的谱图。

线性组分分析土壤中的砷形态，结果主要为 As(Ⅴ)（大于 98%），As(Ⅲ)未被检测到（见表 4.4）。Tang 等人（2016）也测得石门雄黄矿区的土壤中 86%的砷以 As(Ⅴ)形态存在。Button 等人（2011，2012）也发现土壤中的砷形态主要为无机砷，As(Ⅴ)占总提取砷的比例大于 90%。无机砷比其他砷形态溶解性更大、毒性更强，因此，对本研究区的砷污染土壤的修复势在必行。

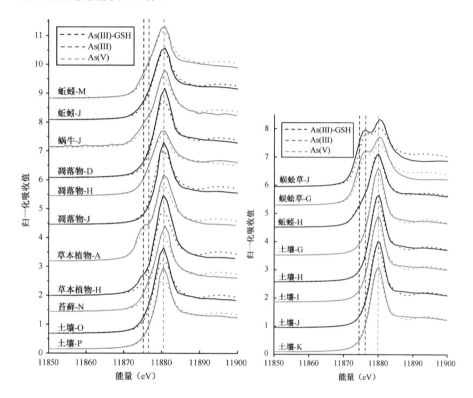

图 4.3 XANES 归一化处理谱（左：固探模式；右：荧光模式）

XANES spectra for model As species 和 selected samples

表 4.4 XANES 分析陆地系统样品中的砷形态

XANES fitting results for selected samples

样品	采样点	砷形态比例（%）			卡方检验
		As(III)-GSH	As(III)	As(V)	
土壤	O	—	—	100	0.0280
	P	—	—	100	0.0134
	G	0.6	—	99.4	0.0030
	H	—	—	100	0.0052

续表

样品	采样点	砷形态比例（%）			卡方检验
		As(Ⅲ)-GSH	As(Ⅲ)	As(Ⅴ)	
	I	0.6	—	99.4	0.0053
	J	1	—	99.0	0.0044
	K	1.7	—	98.3	0.0031
蚯蚓	H	12.7	8.6	78.7	0.0008
	M	6.9	5.3	87.8	0.0076
	J	13.9	—	86.1	0.0066
蜗牛	J	9.8	0.8	89.4	0.0289
凋落物	D	—	—	100	0.0514
	H	—	31.5	68.5	0.0059
	J	0.6	—	99.4	0.0244
草本植物	A	15.7	3.1	81.2	0.0303
	H	0.3	—	99.7	0.0649
苔藓	N	—	3.2	96.8	0.0183
蜈蚣草	J	0.5	65	34.5	0.0192
	G	0.7	58.3	41	0.0180

注：—未检出。

4.3.2.2 植物、凋落物中砷的形态的分布特征

在污染区和非污染区，植物体内的砷形态特点一致。HPLC-ICP-MS 测定的陆地生物中的主要砷形态如表 4.5 所示。结果表明，在大多数情况下，

表 4.5 HPLC-ICP-MS 测定的陆地生物中的主要砷形态

The arsenic species in terrestrial samples determined by HPLC-ICP-MS

样品	采样点	组织	砷形态比例（%）						提取率（%）	柱回收率（%）
			AsC	AsB	未知砷	As(Ⅲ)	DMA	MMA	As(Ⅴ)		
蚯蚓	H					53.20			46.80		
	M					51.98			48.02	34	90
	J			2.55		52.59			44.86	39	110
	L			8.99		52.52			38.49	31	109
蜗牛	J		0.81		tr	72.62	3.28	2.33	20.96	50	114

续表

样品	采样点	组织	砷形态比例（%）							提取率（%）	柱回收率（%）
			AsC	AsB	未知砷	As(III)	DMA	MMA	As(V)		
凋落物	C		9.27		4.90	tr	5.93	9.55	70.35	54	101
	I		4.99		6.96	2.81	7.93	22.47	54.84	69	109
	B		0.45	tr		24.25	33.92	tr	41.38	53	104
	J		0.32		2.75	3.67	7.50	2.33	83.43	49	92
草本植物	I		0.30		1.44	11.22	23.13	1.65	62.26	28	71
	C		2.67		0.85	14.05	3.87	0.95	77.61	42	89
	B		4.95		tr	7.56	3.55	2.38	81.56	68	112
苔藓	N				tr	83.52	7.59		8.89	41	121
苦瓜	M	果实	28.42			71.58			tr	40	55
	M	籽	0.24			5.58			94.18	60	65
茄子	M	叶	3.03		0.38	9.18	0.82	tr	86.59	47	65
黄豆	M	叶	5.77		4.82	34.43	5.12		49.86	89	71
辣椒	L	茎	2.60				6.69		90.71	35	98
玉米	K	叶	2.11		13.97	4.08	16.43	tr	63.41	73	97
玉米	K	茎	3.69		13.86	4.14	5.80		72.51	51	79
南瓜	E	叶	1.43		6.04	42.97	23.34		26.22	49	73
橘子	E	茎	8.60			59.70	17.07	tr	14.63	47	65
橘子	E	叶	2.31		1.63	32.54	34.46	4.73	24.33	59	69
橘子	E	果实	2.86			53.99	6.57		36.58	46	72
芒草	J	叶	4.22		3.89	14.32	10.55	6.62	60.40	44	67
	J	根	0.92		0.24	3.21	1.33	1.95	92.35	36	63
小飞蓬	G	叶	22.93		6.11	3.74	12.11	16.61	38.50	64	105
	G	茎	tr	tr		33.80	27.48		38.72	53	105
蜈蚣草	J	叶				55.48			44.52	55	87
	J	茎				54.76			45.24	58	114
	G	叶	0.51			27.35			72.14	55	96
大叶井口边草	K	叶				51.44			48.56	40	75

注：tr 表示砷形态在色谱中有峰，但是未达到检测限。

无机砷是主要的砷形态，另外还含有少量的有机砷，如 AsC、DMA、MMA 和未知砷。本研究结果与前人研究结果一致（Ruiz-Chancho et al.，2008；Jedynak et al.，2009；Bergqvist and Greger，2012；Larios et al.，2012；Wei et al.，2015）。但对超富集植物而言，如蜈蚣草（茎、叶）和大叶井口边草（叶），HPLC-ICP-MS 分析发现，它们的砷形态基本上只含有无机砷（大于99%），与 XANES 的结果一致，该结果也与前人的研究结果一致（Ma et al.，2001；Lombi et al.，2002；Wang et al.，2002；Wei et al.，2015）。

大多数植物通过凋落物的方式将体内的砷释放到土壤中（Kim et al.，2009）。因此，植物叶片中的砷形态也能表征其相应凋落物中的砷形态特征。在本研究中，污染区和非污染区的植物和草本植物中的砷形态特征与凋落物中的砷形态特征一致，正好说明了这一点。

污染区和非污染区的蚯蚓体内的砷形态主要为无机砷。这一研究与前人关于蚯蚓砷形态的研究结果一致（Langdon et al.，2003a；Watts et al.，2008；Button et al.，2012；Wang et al.，2016）。蚯蚓体内的 As(III) 的比例为 51%～72%，远远高于陆地生态系统中的其他介质，说明蚯蚓也有把 As(V) 还原为 As(III) 的能力。AsB 是蚯蚓体内唯一被检测到的有机砷形态，但是其含量相对较低。

图 4.4 所示为 4 种生物样品的 HPLC-ICP-MS 色谱。

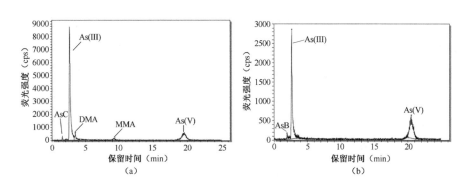

图 4.4 4 种生物样品的 HPLC-ICP-MS 色谱

(a: 蜗牛；b: 蚯蚓；c: 蜈蚣草；d: 玉米)

HPLC chromatograms of the methanol-water-extractable As compounds

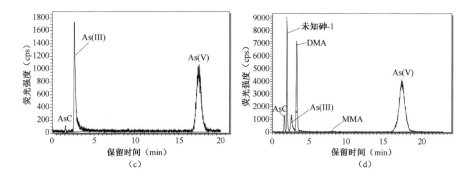

图 4.4 4 种生物样品的 HPLC-ICP-MS 色谱（续）

（a：蜗牛；b：蚯蚓；c：蜈蚣草；d：玉米）

HPLC chromatograms of the methanol-water-extractable As compounds

4.3.3 鸟类中砷的生物富集与转化特征

4.3.3.1 鸟类中砷的分布特征

6 只野生鸟类不同组织中的总砷含量及其砷形态如表 4.6 所示。其中，肌肉中的砷含量范围为 1.13～4.95 mg/kg，高于 Roselli 等人（2016）报道的欧洲 14 种迁徙鸟肌肉中的砷含量（<0.326 mg/kg）。雀形目鸟不同组织（肌肉、心脏、肝脏和胃）中的砷含量一般<1 mg/kg（Sánchez-Virosta et al.，2015）。Dauwe 等人（2005）研究发现，比利时重金属尾矿区的大山雀（*Parus major*）体内砷含量最高也仅为 0.228 mg/kg。Deng 等人（2007）也报道了北京西山的大山雀（*Parus major*）和金翅雀（*Carduelis sinica*）体内砷含量<0.25 mg/kg。在本研究中，鸟类肌肉中的砷含量相对较高说明石门雄黄矿长期的重金属（砷）污染导致鸟类不同程度富集砷。

表4.6 6只野生鸟类不同组织中的总砷含量及其砷形态

HPLC-ICP-MS data: total As concentration, As compounds identified 和 extraction efficiency in bird tissues

鸟类	组织	总砷 (mg/kg)	砷形态比例（%）						提取率（%）	柱回收率（%）
			AsC	AsB	As(III)	DMA	MMA	As(V)		
画眉	肌肉	1.41		5.22	23.64	54.59		16.56	17	66
	羽毛	9.36			36.94	2.32		60.73	53	100
麻雀1	肌肉	3.03			48.10	17.55	2.43	31.92	28	70
	羽毛	49.00			29.24	0.36		70.40	93	88
麻雀2	肌肉	1.75			40.59	29.20		30.21	15	87
	羽毛	51.65			62.69			37.31	54	106
白头翁1	肌肉	1.21			27.83	28.70		43.47	19	50
	羽毛	5.19			31.73	11.54		56.73	28	74
白头翁2	肌肉	1.13			15.76	84.24			26	86
	羽毛	22.68	0.46		6.33	68.75	0.95	23.51	67	109
	爪	2.66			22.41	77.59		tr	66	117
喜鹊	肌肉	4.95		0.72	1.54	97.74	tr	tr	16	90
	羽毛	25.73			11.33	30.44	tr	58.23	76	115
	爪	8.92			5.62	94.38	tr	tr	36	111
	胃	12.16		0.43	9.72	84.60	3.17	2.08	19	120
	胃内容物	135	0.15	0.37	42.14	5.41	2.24	49.69	101	119
	心脏	4.94			9.73	87.44	2.83	tr	11	105
	肝脏	27.82		1.08	14.66	77.11	6.17	0.98	29	87

在本研究中，鸟类羽毛中的砷含量远远高于肌肉中的砷含量，范围为5.19~51.65 mg/kg，表明羽毛跟人类的指甲和头发一样，都是潜在的储存/排泄点位（Matschullat，2000）。羽毛中的砷含量也远远高于前人报道的非污染区鸟类羽毛中的砷含量（一般<3 mg/kg）。Tsipoura 等人（2008）发现，新泽西州北部 Meadowlands District 的雀形目鸟，如红翅黑鹂（*Agelaius phoeniceus*）、马什鹪鹩（*Cistothorus palustris*）、树雀（*Tachycineta bicolour*）的羽毛中的砷含量均<0.2 mg/kg。Borghesi 等人（2016）也发现，西地中海

地区中大火烈鸟（*Phoenicopterus roseus*）羽毛中的砷含量<0.9 mg/kg。Dauwe 等人（2003）发现比利时法兰德斯的中雀鹰（*Accipiter nisus*）、猫头鹰（*Athene nocta*）和仓鸮（*Tyto alba*）等的羽毛中的砷含量<1.40 mg/kg。Cooper 等人（2017）也发现，美国南卡罗莱纳州的大冠蝇霸鹟（*Myiarchus crinitus*）和北美红雀（*Cardinalis cardinalis*）的羽毛中的砷含量分别为 0.175±0.186mg/kg 和 0.117±0.140mg/kg。Tsipoura 等人（2017）报道，新泽西特拉华湾的红腹滨鹬（*Calidris canutus*）、半蹼滨鹬（*Calidris pusilla*）和三趾滨鹬（*Calidris alba*）中的砷含量<2.5 mg/kg。但是在本研究中，鸟类羽毛中的砷含量与 Janssens 等人（2001）的研究结果一致，该研究发现，比利时安特卫普的大山雀（*Parus major*）的羽毛中的砷含量可达到 30.8 mg/kg。这表明羽毛中的砷可以作为有效反映鸟类生存环境中的砷污染水平的重要标志物。尽管在有些情况下，鸟类羽毛并没有清洗，但是 Kim 和 Oh 等人（2014）研究发现，鸟类羽毛表面的污染物无法被清洗完全。因此，鸟类羽毛中的砷可以高度反映环境中的砷污染状况。除了肌肉和羽毛，喜鹊的胃内容物和肝脏中的砷含量也极高，表明鸟类的砷含量来源主要为摄食途径，其胃内容物中的砷含量可达到 136 mg/kg。

4.3.3.2　鸟类中砷形态的分布特征

HPLC-ICP-MS 分析发现，鸟类的砷形态主要为 As(Ⅲ)、DMA 和 As(Ⅴ)，另外还含有少量的 AsB 和 AsC。Gress 等人（2016）也发现，动物园的猎鸟等鸟类体内的砷形态主要为无机砷和 DMA。在本研究中，除白头翁 2 的羽毛中的砷形态以 DMA 为主外，其他鸟类羽毛中的砷形态均以无机砷 As(Ⅴ)和 As(Ⅲ)为主，占提取砷的 69%～100%。研究发现，羽毛中污染物水平可反映鸟类的生长环境的质量（Burger and Gochfeld，2009；García-Fernández et al.，2013；Borghesi et al.，2016）。在本研究中，羽毛中的无机砷含量极高，表明尽管石门雄黄矿关闭十几年，但是当地生物砷暴露依旧相当严重。

选取砷含量相对高（4.95～135.90 mg/kg）的鸟类组织进行 XANES 砷形态分析（见表 4.7）。野生鸟类和家养母鸡的 XANES 归一化处理谱图（实线表示原谱图，虚线表示拟合曲线）如图 4.5 所示。鸟类主要的砷形态为 As(Ⅴ)和 DMA，其次为 As(Ⅲ)（<10.1%）。白头翁 2 羽毛中的 As(Ⅴ)的比例为 41.2%，远高于麻雀 2 和喜鹊羽毛中 As(Ⅴ)的比例。喜鹊肌肉中的砷形态主要为 DMA，但是胃内容物中并没有检测到 DMA，这与 HPLC-ICP-MS 的结果一致。从 XANES 的结果看，As(Ⅲ)-GSH 的含量较高，但是在 HPLC-ICP-MS 中并未被检测到。因此，与 HPLC-ICP-MS 的结果相比，XANES 的结果中 DMA、As(Ⅲ)和 As(Ⅴ)的比例相对较低。As(Ⅲ)-GSH 在大量研究中均有发现（Moriarty et al.，2009；Hong et al.，2014；Foust Jr. et al.，2016）。Ali 等人（2009）曾报道，As(Ⅲ)具有亲巯基（-SH），因此能被绑定到富含巯基的螯合剂上，如谷胱甘肽和植物络合素等。因此，As(Ⅲ)-GSH 可认为是 HPLC-ICP-MS 中未被提取和未检测到的砷形态。

表 4.7 XANES 分析鸟类的砷形态

Table 4.7 XANES fitting results for selected samples in wild birds

样品	组织	砷形态比例（%）				卡方检验
		As(Ⅲ)-GSH	As(Ⅲ)	As(Ⅴ)	DMA	
白头翁 2	羽毛	43.0	0	41.2	15.8	0.0102
麻雀 2	羽毛	75.7	0	11.5	12.8	0.0025
喜鹊	羽毛	77.5	0	16.9	5.6	0.0043
	鸟爪	65.9	0	7.8	26.3	0.0013
	胃内容物	33.9	0	66.1	0	0.0093
	肝脏	71.6	10.1	0	18.3	0.0017
	肌肉	25.0	0	0	75.0	0.0072
家养母鸡-GY2	鸡胗皮	68.2	0	16.6	15.2	0.012
	鸡胗	41.9	5.1	1.2	51.8	0.001
	胃内容物	3.3	0	96.7	0	0.040
	羽毛	10.4	0	65.6	24.0	0.009

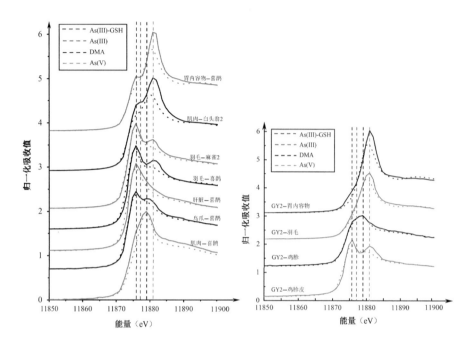

图 4.5　野生鸟类和家养母鸡的 XANES 归一化处理谱图
（实线表示原谱图，虚线表示拟合曲线）

XANES raw data (—) and fitting (---) for model As species and samples

目前，关于陆地生物中 AsB 和 AsC 的形成机理并不明晰（Moriarty et al.，2009）。本研究对喜鹊不同组织中的砷形态分析发现，内脏软组织（肌肉、肝脏、心脏和胃）及鸟爪中最主要的砷形态为 DMA（>77%）。这一结果与一些前人的研究结果相反。Kubota 等人（2003）研究发现，黑尾鸥（*Larus crassirostris*）、黑脚信天翁（*Diomedea nigripes*）和巨嘴鸦（*Corvus macrohynchos*）的肝脏中大于 88%的提取砷均为 AsB。Koch 等人（2005）也发现，枞镰翅鸡（*Dendragapus canadensis*）和灰噪鸦（*Perisoreus canadensis*）的肝脏和肌肉中的砷形态也主要为 AsB。这一差异可能与摄食来源有关，本研究中雀形目鸟主要食物为植物籽粒和昆虫，而鸥类和鸦类主要以动物为食。

4.3.3.4 鸟类砷富集与转化的种间差异

稳定同位素（$\delta^{15}N$ 和 $\delta^{13}C$）常被用来分析食物链中生物的营养级和食源。$\delta^{15}N$ 与食性相关,通常最高营养级生物的 $\delta^{15}N$ 最高(Fry,1988;Watanabe et al.,2008)。$\delta^{13}C$ 反映食物的碳源,不同食物来源的 $\delta^{13}C$ 差别显著(Rounick et al.,1982；Junger and Planas,1994)。根据 $\delta^{15}N$ 判断,本研究中雀形目鸟基本在同一营养级（除麻雀 2 外,$\delta^{15}N$ 的变化范围从 8.01‰~9.57‰）。且本研究中 4 种鸟类食物来源基本一致,均以植物籽粒和昆虫为食。这也可从 $\delta^{13}C$ 的变化范围看出。在本研究中,鸟类的砷含量在同一水平,同时砷形态也基本一致,因此雀形目鸟体内的砷富集和砷转化不存在种间差异。但是,由于本研究数据有限,有待进一步研究。

4.3.4 家禽母鸡中砷的生物富集与转化特征

4.3.4.1 家禽母鸡中砷的分布特征

在本研究中,两处农家后院土壤中的砷含量基本一致,GY 处土壤中的砷含量为 144.31 mg/kg,CY 处土壤中的砷含量为 142.45 mg/kg,远远高于我国土壤环境质量标准中Ⅲ类土壤中的砷含量标准（小于 40 mg/kg）。值得注意的是,CY 位置紧挨着的黄水溪中的水砷含量高达 2629±351 μg/L,是我国饮用水砷标准限值的约 262 倍。说明过去的采矿活动对当地环境造成了严重的砷污染。

家养母鸡不同组织中的总砷及其砷形态如表 4.8 所示。GY 处的家养母鸡不同组织中砷含量从大到小依次为胃内容物（12.99~20.12 mg/kg）、鸡胗皮（8.13～10.58 mg/kg）、羽毛（7.05～7.48 mg/kg）、鸡胗（2.58～3.28 mg/kg）、肝脏（1.35～1.63 mg/kg）、心脏（1.17～1.72 mg/kg）、肌肉（1.02～1.03 mg/kg）和鸡卵（0.73～0.84 mg/kg）。CY 处的家养母鸡不同组织中砷含量表现为羽毛（8.52～15.80 mg/kg）、鸡胗（1.83～2.42 mg/kg）、鸡胗皮（1.99 mg/kg）、肝脏（1.44～1.54 mg/kg）、肌肉（0.82～1.23 mg/kg）和鸡卵（0.88 mg/kg）。

表 4.8 家养母鸡不同组织中的总砷及其砷形态

Total As 和 its species in chicken tissues from Hunan province, China

组织	家养母鸡编号	总砷（mg/kg）（平均值）	砷形态比例（%）						提取率（%）	柱回收率（%）
			AsB	未知1	As(III)	DMA	MMA	As(V)		
肌肉	GY1	1.02	3.86			96.14			19	85
	GY2	1.03	5.75			94.25			62	107
	CY1	1.23	3.12			96.88			49	81
	CY2	0.82	tr			100			24	84
羽毛	GY1	7.48			2.34	9.55	73.91	tr	14.20	41	105
	GY2	7.05		13.90	3.75	60.41	1.45	20.49	28	110
	CY1	15.80	7.49		13.43	11.62		67.46	72	100
	CY2	8.52		5.83	14.01	33.16		47.00	89	95
鸡胗	GY1	3.28	2.40		2.83	94.77			15	102
	GY2	2.58	0.38		1.46	98.16			10	112
	CY1	1.83	tr			100			59	100
	CY2	2.42				100			22	103
鸡胗皮	GY1	10.58			57.10	17.59	11.91	13.40	48	96
	GY2	8.13			27.14	21.93	46.63	4.30	63	106
	CY1	1.99		1.79	67.21	10.06	20.94	tr	27	99
胃内容物	GY1	20.12		1.11	38.02	5.23	11.62	44.02	79	109
	GY2	12.99		1.26	34.64	9.30	39.66	15.14	89	90
心脏	GY1	1.72	5.85		7.09	87.06			14	93
	GY2	1.17	1.65		tr	98.35			14	94

续表

组织	家养母鸡编号	总砷(mg/kg)(平均值)	砷形态比例（%）						提取率(%)	柱回收率(%)
			AsB	未知1	As(III)	DMA	MMA	As(V)		
肝脏	GY1	1.63	4.04		5.13	86.63	4.20		12	110
	GY2	1.35	1.61	1.93	3.28	93.18	tr		13	110
	CY1	1.44			23.48	76.52	tr	tr	77	56
	CY2	1.54	tr		19.61	80.39			68	80
鸡卵	GY1	0.84	tr		tr	100			13	72
	GY2	0.73				100			24	105
	CY2	0.88				100			34	78

在本研究中，家养母鸡羽毛中的砷含量远远高于大多数研究中报道的砷含量，说明与人类头发一样，羽毛也是砷在动物体内的潜在储存/排出点位（Matschullat，2000）。值得一提的是，GY1家养母鸡的鸡胗皮中的砷含量最高，达到 10.58 mg/kg，作为医学使用可能存在一定的健康风险。家养母鸡中最低砷含量为肌肉（0.82 mg/kg），但是高于 Sánchez-Rodas 等人（2006）报道的关于西班牙市场上购买的鸡中的砷含量（0.27 mg/kg）和 Nachman 等人（2013）报道的美国商场购买的鸡中的砷含量（0.003~0.009 μg/kg），以及 Lasky 等人（2004）报道的 1989—2000 年美国农业部食品安全检验局国家残留项目中鸡体内的砷含量（0.10~0.47 mg/kg）。在本研究中，鸡心脏中的砷含量也高于 Pizarro 等人（2015）报道的关于智利商场中购买的鸡心脏中的砷含量（0.06~0.07 mg/kg）。同样，在本研究，鸡肝脏中的砷含量也远远高于 He 等人（2016）报道的鸡肝脏中的砷含量（0.031 mg/kg）。另外，Hu 等人（2017）曾报道我国鸡胗、鸡心脏、鸡肝脏和肌肉中的砷含量小于 1 mg/kg（湿重）。除可食用肉类外，由于家养母鸡生产的鸡蛋营养价值高、无公害而受到消费者的喜爱（Van Eijkeren et al.，2006）。在本研究中，鸡卵中的砷含量换算成湿重即为 0.17~0.22 mg/kg。而在 Grace 和 Macfarlane（2016）的研究中，当土壤中的砷含量小于 18.2 mg/kg

时，鸡卵中的砷含量为0.01～0.06 mg/kg（湿重），远低于本研究中鸡卵中的砷含量。由此可见，环境中的砷污染会导致鸡不同组织中砷的富集。

但是，也有大量研究报道的鸡组织中的砷含量更高。Shah等人（2009b）研究发现，巴基斯坦市场上鸡的肌肉、肝脏和心脏中的砷含量分别为1.70～5.28mg/kg、2.48～7.17mg/kg 和 2.24～6.36 mg/kg。Zhang 等人（2015）也发现我国市场上的鸡肝脏中的砷含量为1.590～5.077 mg/kg。同样的，Kazi等人（2013）研究发现，巴基斯坦农场中养殖的鸡的食物中的砷含量为21.3～43.7 mg/kg，其肌肉、肝脏和心脏中的砷含量分别为2.15～5.28mg/kg，3.07～7.17mg/kg 和 2.11～6.36 mg/kg。另外，He等人（2016）通过室内研究发现，当食物中砷含量为30 mg/kg，90天后鸡肝脏中的砷含量可达到 1.07 mg/kg。可以推断，有些使用含砷饲料的商业鸡肉可能是消费者体内重要的砷来源，因此需要采取相关措施保护公众健康。

4.3.4.2 家养母鸡中砷形态的分布特征

我们在两处农家后院采集的家养母鸡的肌肉、心脏、鸡胗、肝脏中DMA是最主要的砷形态（>76%）。As(Ⅲ)占有机砷的比例<23%，另外还含有少量的其他砷形态，如AsB、MMA、As(Ⅴ)（<5.85%）。这与XANES结果一致，鸡胗中的砷形态主要为DMA（51.8%）。Sanz等人（2005）的研究结果也表示，鸡组织中95%的砷形态为DMA，而AsB和As(Ⅲ)的含量极低。Mao等人（2011）也发现，我国鸡肝脏中DMA和As(Ⅲ)的砷含量比例分别为73%和26%。Hu等人（2017）研究发现，鸡胗中的砷形态主要为DMA和As(Ⅲ)。Pizarro等人（2015）发现，智利鸡心脏中的砷形态主要为As(Ⅴ)、MMA、DMA，其含量分别为 5～8μg/kg、4～8μg/kg、10～16μg/kg。但是也有研究发现，鸡组织中的砷含量主要为AsB。Peng等人（2014）发现，鸡肝脏中的AsB含量为46 μg/kg，远远高于其他砷形态。Liu等人（2015）的研究也发现，鸡组织中的砷形态主要为AsB和As(Ⅲ)，

其含量分别为 101 μg/kg 和 120 μg/kg。另外，大量研究表明，使用含砷饲料的农场中的鸡组织中被检测到大量的 Rox、p-ASA 和苯胂酸。比如，Zhang 等人（2015）报道，鸡肝脏和肌肉中的砷形态主要为 ROX，含量分别为 0.11~2.3 mg/kg、0.15mg/kg。Sánchez-Rodas 等人（2006）也发现，鸡组织中苯胂酸的含量为 227 μg/kg，DMA 基本没有被检测到。Peng 等人（2014）研究发现，使用含 ROX 饲料，鸡肝脏中 ROX 的含量为 151 μg/kg。由此说明，食物来源是鸡体内砷形态不同的主要原因之一。

除可食用肉类组织（肌肉、鸡胗、心脏、肝脏）外，其他组织（鸡卵、鸡胗皮、胃内容物）中的砷形态分析结果如表 4.8 所示。鸡卵中的砷形态为 DMA，可占提取砷含量的 100%，同时微量的 AsB 和 As(Ⅲ)被检测出。GY 处的鸡的羽毛中 DMA 为主要的砷形态；而 CY 处的鸡的羽毛中 As(Ⅴ)为主要的砷形态，这与 XANES 结果一致。在鸡胗皮中，砷形态主要为 As(Ⅲ)、MMA 和 DMA。GY 处的鸡的胃内容物中的砷形态主要为 As(Ⅲ)、As(Ⅴ)和 MMA。这与 XANES 结果也一致。

总的来看，无机砷是羽毛、鸡胗皮和胃内容物中主要的砷形态，但是肌肉、肝脏、心脏、鸡胗和鸡卵中主要为 DMA。

胃内容物是家养母鸡短期储存食物的场所，羽毛中的污染物则可反映周围环境（Burger and Gochfeld，2009；García-Fernández et al.，2013）。因此，可推断鸡自身对摄取的无机砷在体内具有一定的转化作用，形成以甲基砷为主的有机砷。

4.3.4.3 家养母鸡砷富集与转化的差异性

在本研究的两处农家后院中，鸡组织中 $\delta^{15}N$ 的范围为 6.60‰~7.78‰，基本在同一营养级上。但是$\delta^{13}C$ 则表示出极大的差异性，GY 处的$\delta^{13}C$ 为-15.16‰~-13.98‰，而 CY 处的$\delta^{13}C$ 为-20.67‰~-19.52‰。说明在两处农家后院中，鸡的食物来源差异显著。首先，两处土壤中的 pH

值、TOM 和土壤湿度不同,且两处地理环境差异显著。两处鸡的食物习性不同,除大米、玉米等食物外,GY 处的鸡还食用小山丘上生长的植物和砂砾。因此,食物来源是造成两处鸡组织中砷特征差异的原因。但是,很多研究发现,饲养的时间长短、性别等也会影响砷分布特征(Lasky et al.,2004; Liu et al., 2016)。

4.3.5 陆地生态系统中砷的转化

土壤作为陆地生态系统中最重要的砷来源,其上生存的生物中的砷含量远低于土壤中的砷含量。在砷污染区和非污染区,植物体内的砷主要为无机砷,另外还含有少量的有机砷,如 AsC、DMA、MMA 和未知砷。

在本研究中,XANES 的结果表明 As(V)为土壤中最主要的砷形态,但是 As(III)却是某些植物中主要的砷形态。由此说明,植物从土壤中吸收 As(V),并在体内还原为 As(III)。研究表明,植物将 As(V)还原为 As(III)是自身免受 As(V)和磷酸干扰而进行自我保护的有效机制(Wu et al., 2002; Finnegan and Chen, 2012; Pizarro et al., 2016)。

值得一提的是,植物叶片中 AsC 可占提取砷含量的 23%,并且大量未知砷被检测到,说明植物能够直接吸收土壤中的砷并将其转化为更为复杂的砷形态。

有文献报道,在砷污染区和非污染区的土壤中检测到少量的 DMA 和 MMA 等有机砷(Huang and Matner, 2007; Huang et al., 2010; Wang et al., 2016)。说明土壤中复杂的有机砷化合物尽管含量很少,也可能被植物直接吸收富集。但是到目前为止,关于植物体内有机砷是其自身转化作用形成的还是直接从土壤中吸收的,仍不确定。

土壤和凋落物是蚯蚓体内砷的主要来源(Button et al., 2011)。在砷污染区和非污染区，蚯蚓体内的砷形态主要为无机砷，还含有少量 AsB。AsB 的来源可能是直接吸收其摄入的食物或者自身对其他砷形态的转化（Geiszinger et al.，1998；Langdon et al.，2003）。但是，到目前为止，蚯蚓体内 AsB 的来源并没有结论性的证据（Button et al.，2009，2011，2012）。在本研究中，植物和凋落物中并没有检测到 AsB，或者检测到微量的 AsB，但检测到其他有机砷，如 AsC、DMA、MMA 和未知砷，含量相对较高。因此，蚯蚓体内 AsB 不大可能从周围环境中直接获取，有可能是因为蚯蚓自身的转化作用（Langdon et al.，2003；Button et al.，2009）。另外，大量研究表明，环境中的微生物在砷的生物地球化学循环中起着重要作用，包括富集无机砷，转化为甲基砷或其他复杂的有机砷，并释放这些砷形态到环境中（Azizur Rahman et al.，2012；Zeng et al.，2015）。因此，微生物对砷的转化作用对陆地生态系统中生物和环境中的有机砷含量也有一定贡献。从 XANES 结果来看，土壤和蚯蚓中的砷形态组成基本一致。因此，直接从土壤中吸收砷也可能是蚯蚓体内 AsB 的来源。除了小分子砷形态（如无机砷、MMA 和 DMA），蚯蚓可能直接从周围环境中吸收复杂的有机砷作为 AsB 的来源。

植物和土壤动物位于陆地生态系统食物链的底部，是食物链顶端高营养级生物，如鸟类和鸡等的潜在食物来源。不同介质中的砷含量基本表现为：土壤>蚯蚓>植物、凋落物>高级生物鸟类、家养母鸡。在本研究中，土壤中的砷形态基本为无机砷（>98%）；分解者蚯蚓体内主要为无机砷（>91%）；初级生产者（植物）中的砷形态主要为无机砷，占提取砷的 41%~93%；鸟类体内无机砷占 2%~79%；而家养母鸡中基本上并未检测到无机砷。由此可见，无机砷比例随着营养级的提高呈现明显降低趋势。同样，对比不同介质中甲基砷（DMA+MMA）比例发现：土壤（未检测到）≈蚯蚓（未检测到）<植物（1%~38%）<鸟类（17%~97%）<家养母鸡（94%~100%）。因此，不难推断，不同营养级生物对砷的转化能力不同，高营养级生物对无机砷的转化能力明显高于低营养级生物。

4.4 本章小结

本章通过系统研究陆地生态系统中土壤—植物—凋落物—蚯蚓—高级生物对砷的富集特征及其对砷的转化，结果表明生物体对砷的富集作用基本表现为随着土壤中砷含量的增加而增加。污染区土壤中的砷含量为 61~2224 mg/kg，远高于非污染区土壤中的砷含量（8~23 mg/kg）。蚯蚓富集砷的能力远高于草本植物和凋落物，最高可达 430 mg/kg。不同植物对砷的富集能力不同，除超富集植物蜈蚣草和大叶井口边草外，其他植物的砷含量均<16.71 mg/kg。4 种雀形目鸟，包括麻雀、画眉、白头翁和喜鹊肌肉中的砷含量为 1.13~4.95 mg/kg。两处散养鸡肌肉中的砷含量为 0.82~1.23 mg/kg。总的来说，砷含量表现为：土壤>蚯蚓>植物、凋落物>高级生物鸟类、家养母鸡。在陆地生态系统中，除土壤中的砷基本为无机砷外，随着生物营养级的提高，生物体中有机砷比例明显增加。以上成果为揭示砷在陆地生态系统中富集和转化特征，阐明陆地生物和食物链中有机砷的组成与来源，以及为砷的生态与健康风险评估和污染环境修复提供了有价值的数据和信息。

参 考 文 献

ALI W, ISAYENKOV SV, ZHAO FJ, et al. Arsenite transport in plants[J]. Cell Mol Life Sci, 2009, 66:2329-2339.

AMARAL CDB, NÓBREGA JA, NOGUEIRA ARA. Sample preparation for arsenic speciation in terrestrial plants—a review[J]. Talanta, 2013, 115:291-299.

AZIZUR RAHMAN M, HASEGAWA H, LIM RP. Bioaccumulation, biotransformation and trophic transfer of arsenic in the aquatic food chain[J]. Environ Res, 2012, 116:118-135

BELSKII EA, LUGAS'KOVA NV, KARFIDOVA AA. Reproductive parameters of adult birds and morphophysiological characteristics of chicks in the pied flycatcher (*Ficedula hypoleuca* Pall.) in technogenically polluted habitats[J]. Russ J Ecol, 2005, 36:329-335.

BERG B, MCCLAUGHERTY C. Plant litter: Decomposition, humus formation, carbon sequestration[M]. Springer, New York, 2014.

BERGQVIST C, GREGER M. Arsenic accumulation and speciation in plants from different habitats[J]. Appl Geochem, 2012, 27:615-622.

BORGHESI F, MIGANI F, ANDREOTTI A, et al. Metals and trace elements in feathers: a geochemical approach to avoid misinterpretation of analytical responses[J]. Sci Total Environ, 2016, 544:476-494.

BROOKS PR. Plants that hyperaccumulate heavy metals[M]. CAB International, Wallingford, New York, 1998.

BURGER J, GOCHFELD M. Comparison of arsenic, cadmium, chromium, lead, manganese, mercury and selenium in feathers in bald eagle (*Haliaeetus leucocephalus*), and comparison with common eider (*Somateria mollissima*), glaucous-winged gull (*Larus glaucescens*), pigeon guillemot (*Cepphus columba*), and tufted puffin (*Fratercula cirrhata*) from the Aleutian Chain of Alaska[J]. Environ Monit Assess, 2009, 152:357-367.

BUTTON M, JENKIN GRT, HARRINGTON CF, et al. Arsenic biotransformation in earthworms from contaminated soils[J]. J Environ Monitor, 2009, 11:1484-1191.

BUTTON M, MORIARTY MM, WATTS M, et al. Arsenic speciation in field-collected and laboratory-exposed earthworms *Lumbricus terrestris*[J]. Chemosphere, 2011, 85:1277-1283.

BUTTON M, KOCH I, REIMER KJ. Arsenic resistance and cycling in earthworms residing at a former gold mine in Canada[J]. Environ Pollut, 2012, 169:74-80.

CALISI A, ZACCARELLI N, LIONETTO MG, et al. Integrated biomarker analysis in the earthworm *Lumbricus terrestris*: Application to the monitoring of soil heavy metal pollution[J]. Chemosphere, 2013, 90:2637-2644.

CAO X, MA LQ, TU C. Antioxidative responses to arsenic in the arsenic-hyperaccumulator Chinese brake fern (*Pteris vittata L.*)[J]. Environ Pollut, 2004, 128:317-325.

COOPER Z, BRINGOLF R, COOPER R, et al. Heavy metal bioaccumulation in two passerines with differing migration strategies[J]. Sci Total Environ, 2017, 592, 25-32.

DAUWE T, BERVOETS L, PINXTEN R, et al. Variation of heavy metals within and among feathers of birds of prey: Effects of molt and external contamination[J]. Environ Pollut, 2003, 124:429-436.

DAUWE T, JANSSENS E, BERVOETS L, et al. Heavy-metal

concentrations in female laying great tits (*parus major*) and their clutches[J]. Arch Environ Con Tox, 2005, 49:249-256.

DADA EO, NJOKU KL, OSUNTOKI AA, et al. Evaluation of the response of a wetland, tropical earthworm to heavy metal contaminated soil[J]. Int J Environ Monit Anal, 2013, 1:47-52.

DA SILVA SOUZA T, CHRISTOFOLETTI CA, BOZZATTO V, et al. The use of diplopods in soil ecotoxicology—a review[J]. Ecotox Environ Safe, 2014, 103:68–73.

DEDEKE GA, OWAGBORIAYE FO, ADEBAMBO AO, et al. Earthworm metallothionein production as biomarker of heavy metal pollution in abattoir soil[J]. Appl Soil Ecol, 2016, 104:42-47.

DENG HL, ZHANG ZW, CHANG CY, et al. Trace metal concentration in Great Tit (Parus major) and Greenfinch (Carduelis sinica) at the Western Mountains of Beijing, China[J]. Environ Pollut, 2007, 148:620-626.

EEVA T, AHOLA M, LEHIKOINEN E. Breeding performance of blue tits (*cyanistes caeruleus*) and great tits (*parus major*) in a heavy metal polluted area[J]. Environ Pollut, 2009, 157:3126-3131.

FAOSTAT. Food and Agriculture Organization of the United States. 2014, http://faostat3.fao.org/browse/Q/QA/E.

FINNEGAN PM, CHEN WH. Arsenic toxicity: the effects on plant metabolism[J]. Front Physiol, 2012, 3:182-192.

FOUST JR RD, BAUER AM, COSTANZA-ROBINSON M, et al. Arsenic transfer and biotransformation in a fully characterized freshwater food web[J]. Coordin Chem Rev, 2016, 306:558-565.

FREITAS H, PRASAD MNV, PRATAS J. Plant community tolerant to trace elements growing on the degraded soils of São Domingos mine in the South east of Portugal: Environmental implications[J]. Environ Int, 2004, 30:65-72.

FRY B. Food web structure on Georges Bank from stable C, N, and S isotopic compositions[J]. Limnol. Oceanogr, 1988, 33:1182-1190.

FU Z, WU F, MO C, et al. Bioaccumulation of antimony, arsenic, and mercury in the vicinities of a large antimony mine, China[J]. Microchem J, 2011, 97:12-19.

GARCÍA-FERNÁNDEZ AJ, ESPÍN S, MARTÍNEZ-LÓPEZ E. Feathers as a biomonitoring tool of polyhalogenated compounds: A review[J]. Environ Sci Technol, 2013, 47:3028-3043.

GARCÍA-GÓMEZ C, ESTEBAN E, SÁNCHEZ-PARDO B, et al. Assessing the ecotoxicological effects of longterm contaminated mine soils on plants and earthworms: relevance of soil (total and available) and body concentrations[J]. Ecotoxicology, 2014, 23:1195-1209.

GARELICK H, JONES H, DYBOWSKA A, et al. Arsenic pollution sources[J]. Rev Environ Contam T, 2008, 197:17-60.

GEISZINGER A, GOESSLER W, KUEHNELT D, et al. Determination of arsenic compounds in earthworms[J]. Environ Sci Technol, 1998, 32:2238-2243.

GRACE EJ, MACFARLANE GR. Assessment of the bioaccumulation of metals to chicken eggs from residential backyards[J]. Sci Total Environ, 2016, 563-564:256-260.

GRESS J, DA SILVA EB, DE OLIVEIRA LM, et al. Potential arsenic exposures in 25 species of zoo animals living in cca-wood enclosures[J]. Sci Total Environ, 2016, 551-552:614-621.

HE Y, SUN BN, LI SW, et al. Simultaneous analysis 26 mineral element contents from highly consumed cultured chicken overexposed to arsenic trioxide by inductively coupled plasma mass spectrometry[J]. Environ Sci Pollut R, 2016, 23:21741-21750.

HONG S, KHIM JS, PARK J, et al. Species and tissue-specific bioaccumulation of arsenicals in various aquatic organisms from a highly industrialized area in the Pohang City, Korea[J]. Environmental Pollution, 2014.

HU YN, ZHANG WF, CHENG HF, et al. Public health risk of arsenic species in chicken tissues from live poultry markets of Guangdong province, China[J]. Environ Sci Technol, 2017, 51:3508-3517.

HUANG JH, MATZNER E. Mobile arsenic species in unpolluted and polluted soils[J]. Sci Total Environ, 2007, 377:308-318.

HUANG JH, HU KN, DECKER B. Organic arsenic in the soil environment: Speciation, occurrence, transformation, and adsorption behavior[J]. Water Air Soil Poll, 2010, 219:401-415.

HUGHES MF. Arsenic toxicity and potential mechanisms of action[J]. Toxicol Lett, 2002, 133:1-16.

JANSSENS E, DAUWE T, BERVOETS L, et al. Heavy metals and selenium in feathers of great tits (*Parus major*) along a pollution gradient[J]. Environ Toxicol Chem, 2001, 20:2815–2820.

JANSSENS E, DAUWE T, PINXTEN R, et al. Breeding performance of great tits (*Parus major*) along a gradient of heavy metal pollution[J]. Environ Toxicol Chem, 2003, 22:1140-1145.

JEDYNAK L, KOWALSKA J, HARASIMOWICZ J, et al. Speciation analysis of arsenic in terrestrial plants from arsenic contaminated area[J]. Sci

Total Environ, 2009, 407:945-952.

JEONG S, MOON HS, NAM K. Enhanced uptake and translocation of arsenic in Cretan brake fern (*Pteris cretica L.*) through siderophorearsenic complex formation with an aid of rhizospheric bacterial activity[J]. J Hazard Mater, 2014, 280:536-543.

JONCZAK J, PARZYCH A, SOBISZ Z. Dynamics of Cu, Mn, Ni, Sr and Zn release during decomposition of four types of litter in headwater riparian forests in northern Poland[J]. Leśne Prace Badawcze (*Forest Research Papers*), 2014, 75:193–200.

JUNGER M, PLANAS D. Quantitative use of stable carbon isotope analysis to determine the trophic base of invertebrate communities in a boreal forest lotic system[J]. Can J Fish Aquat Sci, 1994, 51:52-61.

JUSSELME MD, MIAMBI E, MORA P, et al. Increased lead availability and enzyme activities in rootadhering soil of *Lantana camara* during phytoextraction in the presence of earthworms[J]. Sci Total Environ, 2013, 445-446:101-109.

KAZI TG, SHAH AQ, AFRIDI HI, et al. Hazardous impact of organic arsenical compounds in chicken feed on different tissues of broiler chicken and manure[J]. Ecotox Environ Safe, 2013, 87:120-123.

KIM YT, YOON HO, YOON C, et al. Arsenic species in ecosystems affected by arsenic-rich spring water near an abandoned mine in Korea[J]. Environ Pollut, 2009, 157:3495-3501.

KIM J, OH JM. Relationships of metals between feathers and diets of Black-tailed Gull (*Larus crassirostris*) chicks[J]. B Environ Contam Tox, 2014a, 92:265-269.

KIM J, OH JM. Lead and cadmium contaminations in feathers of heron and egret chicks[J]. Environ Monit Assess, 2014b, 186:2321-2327.

KOCH I, MACE JV, REIMER KJ. Arsenic speciation in terrestrial birds from Yellowknife, Northwest Territories, Canada: the unexpected finding of arsenobetaine[J]. Environ Toxicol Chem, 2005, 24:1468-1474.

KRAMAR U, NORRA S, BERNER Z, et al. On the distribution and speciation of arsenic in the soil-plant system of a rice field in West-Bengal, India: A μ-synchrotron techniques based case study[J]. Appl Geochem, 2017, 77:4-14.

KUBOTA R, KUNITO T, TANABE S. Occurrence of severak arsenic compounds in the liver of birds, cetaceans, pinnipeds, and sea turtles[J]. Environ Toxicol Chem, 2003, 22:1200-1207.

LANGDON CJ, PIEARCE TG, FELDMANN J, et al. Arsenic speciation in the earthworms *Lumbricus rubellus* and *Dendrodrilus rubidus*[J]. Environ Toxicol Chem, 2003a, 22:1302-1308.

LANGDON CJ, PIEARCE TG, MEHARG AA, et al. Interactions between earthworms and arsenic in the soil environment: A review[J]. Environ Pollut, 2003b,124:361-373.

LARIOS R, FERNÁNDEZ-MARTÍNEZ R, LEHECHO I, et al. A methodological approach to evaluate arsenic speciation and bioaccumulation in different plant species from two highly polluted mining areas[J]. Sci Total Environ, 2012, 414:600-607.

LASKY T, SUN W, KADRY A, et al. Mean total arsenic concentrations in chicken 1989-2000 and estimated exposures for consumers of chicken[J]. Environ Health Persp, 2004, 112:18-21.

LEE BT, LEE SW, KIM KR, et al. Bioaccumulation and soil factors affecting the uptake of arsenic in earthworm, *Eisenia fetida*[J]. Environ Sci Pollut R, 2013, 20:8326-8333.

LIU XP, ZHANG WF, HU Y, et al. Arsenic pollution of agricultural soils by concentrated animal feeding operations (CAFOs)[J]. Chemosphere, 2015a, 119:273-281.

LIU QQ, PENG HY, LU XF, et al. Enzyme-assisted extraction and liquid chromatography mass spectrometry for the determination of arsenic species in chicken meat[J]. Analytica Chimica Acta, 2015b, 888:1-9.

LIU QQ, PENG HY, LU XF, et al. Arsenic species in chicken breast: temporal variations of metabolites, elimination kinetics, and residual concentrations[J]. Environ Health Persp, 2016, 124:1174-1181.

LOMBI E, ZHAO FJ, FUHRMANN M, et al. Arsenic distribution and speciation in the fronds of the hyperaccumulator *Pteris vittata*[J]. New Phytol, 2002, 156:195-203.

LUCISINE P, LECERF A, DANGER M, et al. Litter chemistry prevails over litter consumers in mediating effects of past steel industry activities on leaf litter decomposition[J]. Sci Total Environ, 2015, 537:213-224.

MA LQ, KOMAR KM, TU C, et al. A fern that hyperaccumulates arsenic[J]. Nature, 2001, 409:579-579.

MANDAL BK, SUZUKI KT. Arsenic round the world: a review. Talanta, 2002, 58:201-235.

MAO XJ, CHEN BB, HUANG CZ, et al. Titania immobilized polypropylene hollow fiber as a disposable coating for stir bar sorptive extraction–high performance liquid chromatographyinductively coupled plasma mass spectrometry speciation of arsenic in chicken tissues[J]. J Chromatogr A,

2011, 1218:1-9.

MATSCHULLAT J. Arsenic in the geosphere-a review[J]. Sci Total Environ, 2000, 249:297-312.

MAUNOURY-DANGER F, FELTEN V, BOJIC C, et al. Metal release from contaminated leaf litter and leachate toxicity for the freshwater crustacean *Gammarus fossarum*[J]. Environ Sci Pollut R, 2017, doi:10.1007//s11356-017-9452-0

MORIARTY MM, KOCH I, GORDON RA, et al. Arsenic speciation of terrestrial invertebrates[J]. Environ Sci Technol, 2009, 43:4818-4823.

NACHMAN KE, BARON PA, GEORG R, et al. Roxarsone, inorganic arsenic, and other arsenic species in chicken: a U.S.-based market basket sample[J]. Environ Health Persp, 2013, 121:818-824.

NG JC. Environmental contamination of arsenic and its toxicological impact on humans[J]. Environ Chem, 2005, 2:146-160.

NIAZI NK, SINGH B, SHAH P. Arsenic speciation and phytoavailability in contaminated soils using a sequential extraction procedure and XANES spectroscopy[J]. Environ Sci Technol, 2011, 45:7135-7142.

OTONES V, ÁLVAREZ-AYUSO E, GARCÍA-SÁNCHEZ A, et al. Arsenic distribution in soils and plants of an arsenic impacted former mining area[J]. Environ Pollut, 2011, 159:2637-2647.

PENG H, HU B, LIU Q, et al. Liquid chromatography combined with atomic and molecular mass spectrometry for speciation of arsenic in chicken liver[J]. J Chromatogr A, 2014, 1370:40-49.

PICZAK K, LEŚNIEWICZ A, ŻYRNICKI W. Metal concentrations in deciduous tree leaves from urban areas in Poland[J]. Environ Monit Assess,

2003, 86:273-287.

PIZARRO I, GÓMEZ M, CÁMARA C, et al. Arsenic speciation in environmental and biological samples extration and stability studies[J]. Analytica Chimica Acta, 2003, 495:85-98.

PIZARRO I, ROMÁN D, GÓMEZ MM, et al. Arsenic status and speciation in chicken heart tissues[J]. J Chil Chem Soc, 2015, 60:2986-2992.

PIZARRO I, GÓMEZ-GÓMEZ M, LEÓN J, et al. Bioaccessibility and arsenic speciation in carrots, beets and quinoa from contaminated area of Chile[J]. Sci Total Environ, 2016, 565:557-563.

PULFORD ID, WATSON C. Phytoremediation of heavy metal contaminated land by trees—a review[J]. Environ Int, 2003, 29:529-540.

ROMERO-FREIRE A, MARTÍN PEINADO FJ, DÍEZ ORTIZ M, et al. Influence of soil properties on the bioaccumulation and effects of arsenic in the earthworm *Eisenia andrei*[J]. Environ Sci Pollut R, 2015, 22:15016-15028.

ROSAS M, GUZMÁN-MAR JL, ALFARO JM, et al. 2014. Evaluation of the transfer of soil arsenic to maize crops in suburban areas of San Luis Potosi, Mexico[J]. Sci Total Environ, 2014, 497-498:153-162.

ROSELLI C, DESIDERI D, ASSUNTA MELI, et al. Essential and toxic elements in meat of wild birds[J]. Journal of Toxicology and Environmental Health, Part A, 2016, 79:1008-1014.

ROUNICK JS, WINTERBOURNE MJ, LYON GL. Differential utilization of allochthonous and autochthonous inputs by aquatic invertebrates in some New Zealand streams: A stable carbon isotope study[J]. Oikos, 1982, 39:191-198.

RUIZ-CHANCHO MJ, LÓPEZ-SÁNCHEZ JF, SCHMEISSER E, et al. Arsenic speciation in plants growing in arsenic-contaminated sites[J].

Chemosphere, 2008, 71:1522-1530.

SÁNCHEZ-RODAS D, LUIS GJ, OLIVEIRA V. Development of a rapid extraction procedure for speciation of arsenic in chicken meat[J]. Anal Bioanal Chem, 2006, 385:1172-1177.

SÁNCHEZ-VIROSTA P, ESPÍN S, GARCÍA-FERNÁNDEZ AJ, et al. A review on exposure and effects of arsenic in passerine birds[J]. Sci Total Environ, 2015, 512-513:506-525.

SANZ E, MUÑOZOLIVAS R, CÁMARA C. Evaluation of a focused sonication probe for arsenic speciation in environmental and biological samples[J]. J Chromatogr A, 2005, 1097:1-8.

SHAH AQ, KAZI TG, ARAIN MB, et al. Comparison of electrothermal and hydride generation atomic absorption spectrometry for the determination of total arsenic in broiler chicken[J]. Food Chem, 2009b, 113:1351-1355.

SHEN H, NIU Q, XU, MC, et al. Factors affecting arsenic methylation in arsenic-exposed humans: A systematic review and meta-analysis[J]. Int J Env Res Pub He, 2016, 13:205-222.

SHIN KH, KIM JY, KIM KW. Earthworm toxicity test for the monitoring arsenic and heavy metal containing mine tailings[J]. Environ Eng Sci, 2007, 24:1257-1265.

SIVAKUMAR S. Effects of metals on earthworm life cycles: A review[J]. Environ Monit Assess, 2015, 187:530-545.

SIZMUR T, PALUMBO-ROE B, WATTS MJ, et al. Impact of the earthworm *Lumbricus terrestris* (L.) on As, Cu, Pb and Zn mobility and speciation in contaminated soils[J]. Environ Pollut, 2011a, 159:742-748.

SIZMUR T, WATTS MJ, BROWN GD, et al. Impact of gut passage and mucus secretion by the earthworm *Lumbricus terrestris* on mobility and speciation of arsenic in contaminated soil[J]. J Hazard Mater, 2011b, 197:169-175.

SMEDLEY PL, KINNIBURGH DG. A review of the source, behaviour and distribution of arsenic in natural waters[J]. Appl Geochem, 2002, 17, 517-568.

TANG JW, LIAO YP, YANG ZH, et al. Characterization of arsenic serious-contaminated soils from Shimen realgar mine area, the Asian largest realgar deposit in China[J]. J Soil Sediment, 2016, 16,1519-1528.

TSIPOURA N, BURGER J, FELTESD R, et al. Metal concentrations in three species of passerine birds breeding in the Hackensack Meadowlands of New Jersey[J]. Environ Res, 2008, 107:218-228.

TSIPOURA N, BURGER J, NILES L, et al. Metal levels in shorebird feathers and blood during migration through Delaware Bay[J]. Arch Environ Con Tox, 2017, 72:562-574.

VAN GESTEL CAM, KOOLHAAS JE, HAMERS T, et al. Effects of metal pollution on earthworm communities in a contaminated floodplain area: Linking biomarker, community and functional responses[J]. Environ Pollut, 2009, 157:895–903.

VAN EIJKEREN JC, ZEILMAKER MJ, KAN CA, et al. A toxicokinetic model for the carry-over of dioxins and PCBs from feed and soil to eggs[J]. Food Addit Contam, 2006, 23:509-517.

VAN NEVEL L, MERTENS J, DEMEY A, et al. Metal and nutrient dynamics in decomposing tree litter on a metal contaminated site[J]. Environ Pollut, 2014, 189:54–62.

WAN X, LEI M, CHEN T, et al. Micro-distribution of arsenic species in tissues of hyperaccumulator *Pteris vittata* L.[J]. Chemosphere, 2017, 166:389-399.

WANG JR, ZHAO FJ, MEHARG AA, et al. Mechanisms of arsenic hyperaccumulation in *Pteris vittata*: Uptake kinetics, interactions with phosphate, and arsenic speciation[J]. Plant Physiol, 2002,130:1552-1561.

WANG ZF, CUI ZJ, LIU L, et al. Toxicological and biochemical responses of the earthworm *Eisenia fetida* exposed to contaminated soil: Effects of arsenic species[J]. Chemosphere, 2016, 154, 161-170.

WATANABE K, MONAGHAN MT, TAKEMON Y, et al. Biodilution of heavy metals in a stream macroinvertebrate food web: Evidence from stable isotope analysis[J]. Sci Total Environ, 2008, 394:57-67.

WATTS MJ, BUTTON M, BREWER TS, et al. Quantitative arsenic speciation in two species of earthworms from a former mine site[J]. J Environ Monitor, 2008, 10:753-759.

WEI CY, SUN X, WANG C, et al. Factors influencing arsenic accumulation by *Pteris vittata*: A comparative field study at two sites[J]. Environ Pollut, 2006, 141:488-493.

WEI CY, GE ZF, CHU W, et al. Speciation of antimony and arsenic in the soils and plants in an old antimony mine[J]. Environ Exp Bot, 2015, 109:31-39.

WU JH, ZHANG R, LILLEY RM. Methylation of arsenic in vitro by cell extracts from bentgrass (*Agrostis tenuis*): Effect of acute exposure of plants to arsenate[J]. Funct Plant Biol, 2002, 29:73-80.

ZENG XB, SU SM, FENG QF, et al. Arsenic speciation transformation and arsenite influx and efflux across the cell membrane of fungi investigated using

HPLC-HG-AFS and in-situ XANES[J]. Chemosphere, 2015, 119:1163-1168.

ZHANG WF, HU YN, CHENG HF. Optimization of microwave-assisted extraction for six inorganic and organic arsenic species in chicken tissues using response surface methodology[J]. J Sep Sci, 2015, 38:3063-3070.

第 5 章

水生环境中砷的生物富集与转化

5.1 水生环境中砷的生物富集与转化研究意义

砷在环境中普遍存在，对动植物有广泛的致毒性（Mandal and Suzuki，2002；Smedley and Kinniburgh，2002；Ng，2005）。水生生物可以通过多种渠道摄入食物、水、沉积物中的重金属，并在体内富集，最终进入人体，引发一系列的健康问题（Dhaneesh et al.，2012；Alamdar et al.，2017）。近年来，水生生态系统中的砷污染受到了国内外的广泛关注。淡水鱼类的砷含量相对较低，一般<1 mg/kg，远低于海洋生物（5~50 mg/kg）（Mandal and Suzuki，2002；Kucuksezgin et al.，2014）。相对于大量有关海洋生物的研究而言，淡水水生生物的砷含量与砷形态特征的报道尚不多见（Azizur Rahman et al.，2012）。

一般研究认为，砷更多地富集于食物链的底端，即生产者和初级消费者中，在高级消费者中砷富集程度趋于降低（Chen and Folt，2000；Culioli et al.，2009；Cui et al.，2011；Revenga et al.，2012；Vizzini et al.，2013；Dovick et al.，2016；Alamdar et al.，2017）。近年来，碳氮同位素比值能够反映研究对象的上级营养关系，是有效分析生态系统中营养级关系的很重要的手段（Fry，1988；Moriarty et al.，2009；Saigo et al.，2015）。应用碳氮同位素分析发现，砷沿着食物链营养级的传递过程中没有生物放大作用（Ikemoto et al.，2008；Watanabe et al.，2008；Pereira et al.，2010；Cui et al.，2011；Revenga et al.，2012；Vizzini et al.，2013；Liu et al.，2018）。Juncos 等人（2016）研究发现，Nahuel 湖食物网中砷存在生物稀释作用。但是，生物对砷的富集受到多种因素的影响，如栖息水层、食性等（Farag et al.，2007；Culioli et al.，2009），因此需要进一步研究。

研究表明，砷的毒性不仅与其总量有关，更取决于其化学形态特征（Mandal and Suzuki，2002）。海洋生物中的砷形态相对比较简单，主要为 AsB，而无机砷形态的比例很小（Mandal and Suzuki，2002；Hirata and Toshimitsu，2007；Maher et al.，2009；Grinham et al.，2014；Krishnakumar et al.，2016）。淡水生物中的砷形态更为复杂和多样。很多研究发现，淡水鱼类中主要为有机砷，但与海洋生物相比，AsB 的含量比例显著下降（Šlejkovec et al.，2004；de Rosemond et al.，2008；Miyashita et al.，2009；Ciardullo et al.，2010；Ruttens et al.，2012；Hong et al.，2014）。不同物种的砷形态差别较大。有研究报道，淡水鱼类中的砷形态主要为 DMA（Jankong et al.，2007；Miyashita et al.，2009；Ruttens et al.，2012；Cott et al.，2016；Yang et al.，2017）。也有研究表明，AsB 是淡水鱼类中的主要砷形态（Shiomi et al.，1995；Šlejkovec et al.，2004；Miyashita et al.，2009；Ciardullo et al.，2010；Ruttens et al.，2012；Hong et al.，2014；Juncos et al.，2019）。到目前为止，人们对于淡水鱼中存在的主要砷形态还没有一个全面的认识（Azizur Rahman et al.，2012）。尽管已有一些对食物链中砷富集传递作用的报道，但尚未关注不同营养级生物体中砷的生物转化行为与富集之间的关系（Maeda et al.，1993；Chen and Folt，2000；Culioli et al.，2009；Mogren et al.，2013）。

食物链中不同营养级的生物对砷的代谢能力和代谢机制有所不同。

Maeda 等人（1992）选用念珠藻→黑壳虾→鲤鱼食物链，发现甲基砷化合物会随着营养级的提高而连续增加。Kuroiwa 等人（1994）发现，生物体中甲基砷随食物链营养级升高而增加。因此，甲基砷是否会在食物链中不断富集，对生物体产生危害乃至影响食物链的结构，需要开展更多的研究加以回答。

石门雄黄矿（29°38′11″～29°38′43″N，111°2′06″～111°2′23″E）位于湖南省石门县，是亚洲最大的砷矿，已经有1500多年的采矿历史了（Zhu et al.，2010）。尽管该矿已经被关闭15年了，但是仍然有大量的尾矿和固体废物污染当地环境，研究发现，土壤中的砷含量可达到5240 mg/kg，水中的砷含量可达到40.1 mg/L（Zhu et al.，2015；Tang et al.，2016）。因此，石门雄黄矿是研究砷的理想地区。本研究的目的是：①水生生物对砷的富集，以及其影响因素；②通过碳氮同位素分析，研究水生生态系统中的生物在食物链传递过程中对砷的富集和转化规律；③当地居民砷摄入的健康风险评估。

5.2 砷矿区水生环境中生物样品采集、总砷及砷形态分析方法

本研究选取区域 2、区域 3、区域 4、区域 5 和区域 7 为采样区域，于 2015 年 5 月和 2016 年 6 月采样。其中，区域 2、区域 3、区域 4、区域 5 的水生生物采集方式为雇用当地渔民电打，区域 7 的水生生物采集方式为渔网捕捞。

水生生物主要包括鲤鱼、鲫鱼、鳙鱼、马口鱼、鳝鱼、泥鳅等鱼类，以及其他水生生物，如青蛙、螃蟹、龙虾等。样品采集后马上记录取样信息，如体长、体重等。所有生物样品解剖后取肌肉（背部），个别大型鱼类取肝脏、鱼皮等组织，清洗后置于自封袋中冷冻保存。所有组织样品在冷冻、干燥和研磨后进行化学分析，化学分析方法见第 2 章。石门雄黄矿区水生生物样品的相关参数特征如表 5.1 所示。

表 5.1 石门雄黄矿区水生生物样品的相关参数特征

Selected parametric characteristics of aquatic organisms

采样点	样品名称	n	体重（g）	体长（cm）	栖息水层[①]	食性[①]	$\delta^{13}C$ （‰）	$\delta^{15}N$ （‰）
区域 2	螃蟹	14	19～27		底栖	杂食	−20.61±0.03	7.20±0.14
	泥鳅	11	3～15	7～15	底栖	杂食	−19.71±0.08	9.11±1.57
	鲫鱼	6	4～8	6～11	中层	杂食	−26.79±1.24	6.92±0.97
	马口鱼	>30	4～14	6～11	上层	杂食	−20.45±2.62	8.89±0.99
	麦穗鱼	5	1～3	4～6	上层	杂食	−23.09±0.09	8.17±0.05
	鲤鱼	5	8～98	8～17	底栖	杂食	−22.08±1.69	7.39±0.35
	宽鳍鱲	>30	13～35	10～13	上层	杂食	−19.66±0.04	8.10±0.11
	翘嘴红鲌	6	5～8	7～11	上层	肉食	−19.34±0.05	7.54±0.03
区域 3	青蛙	3	31～35		两栖动物	肉食	−23.18±0.09	7.46±0.03
	螃蟹	5	20～25		底栖	杂食	−20.15±0.21	5.77±0.19
	泥鳅	13	6～20	10～19	底栖	杂食	−22.97±2.58	6.62±1.08
	鲫鱼	5	2～11	5～8	中层	杂食	−26.41±1.13	6.24±0.41
	马口鱼	>30	2～19	5～10	上层	杂食	−19.52±1.43	9.78±0.67
	麦穗鱼	3	3～4	6～7	上层	杂食	−23.07±0.10	9.37±0.08
	宽鳍鱲	>30	4～9	6～7	上层	杂食	−19.83±0.96	9.33±0.49
	翘嘴红鲌	5	6～8	9～11	上层	肉食	−28.84±0.19	10.47±0.06
区域 4	泥鳅	9	3～8	7～12	底栖	杂食	−16.27±0.34	10.60±0.50
	鲫鱼	5	4～6	7～9	中层	杂食	−21.67±2.28	8.62±1.80
	马口鱼	>30	1～13	4～13	上层	杂食	−17.43±0.78	10.20±0.64
	沙塘鳢	14	2～14	5～11	底栖	肉食	−16.67±0.97	11.58±0.65
	宽鳍鱲	>30	10～19	7～11	上层	杂食	−18.06±0.68	10.61±0.57
	翘嘴红鲌	5	22～28	7～13	上层	肉食	−16.42±0.17	12.77±0.15
区域 5	青蛙	2	38～40		两栖动物	肉食	−23.32±0.02	6.33±0.08
	鲫鱼	7	7～55	3～15	中层	杂食	−18.86±0.07	6.99±0.06
	马口鱼	>30	1～16	5～10	上层	杂食	−17.62±2.07	10.09±0.44
	沙塘鳢	17	9～11	7～8	底栖	肉食	−18.79±2.70	11.16±0.38
	桂花鱼	4	120～130	20～25	底栖	肉食	−16.96±0.12	12.82±0.34
	宽鳍鱲	>30	14～20	9～11	上层	杂食	−18.21±1.43	9.45±0.88
	翘嘴红鲌	7	2～38	3～17	上层	肉食	−27.39±0.70	8.52±0.18

续表

采样点	样品名称	n	体重（g）	体长（cm）	栖息水层[①]	食性[①]	$\delta^{13}C$ (‰)	$\delta^{15}N$ (‰)
区域6	螃蟹	8	18~21		底栖	杂食	-24.68±0.27	8.06±0.25
	龙虾	>30	2~27	5~11	底栖	杂食	-25.94±1.30	9.47±0.78
	青虾	>50	1~4	5~6	底栖	杂食	-28.16±0.59	9.58±1.18
	鲫鱼	16	21~195	5~22	中层	杂食	-24.56±2.31	8.18±0.55
	鳊鱼	9	360~550	33~37	中层	草食	-27.76±1.56	7.45±0.57
	鳙鱼	5	300~1350	28~45	上层	滤食	-24.80±0.83	8.06±0.99
	银鮈	3	4~8	8~9	中层	杂食	-25.47±1.41	9.65±0.40
	蛇鮈	2	34~37	16~18	中层	杂食	-26.70±0.08	9.91±0.15
	黄颡鱼	17	7~205	9~27	中层	杂食	-25.54±1.52	9.83±0.84
	鲶鱼	4	88~500	25~48	中层	肉食	-25.60±1.50	10.78±0.95
	鳝鱼	12	72~120	42~49	底栖	肉食	-24.28±0.04	12.48±0.03
	白鲢	3	410~570	32~36	上层	滤食	-25.86±1.01	5.01±0.98
	桂花鱼	7	16~750	12~35	底栖	肉食	-25.79±0.27	12.48±0.43
	鳑鲏	11	5~146	7~12	上层	杂食	-26.28±0.68	10.61±0.54
	鲤鱼	8	30~1700	12~53	底栖	杂食	-26.92±2.85	6.66±1.66
	翘嘴红鲌	13	15~500	10~43	上层	肉食	-26.83±0.80	9.70±1.15

① 栖息水层和食性的信息主要来自网站 FishBase 和其他文献（Li et al.，2011；Zhang et al.，2013；Zhong et al.，2018；Fan et al.，2019）。

另外，浮游生物的采集方法如下：浮游植物用 25#浮游生物网（网格宽 0.064 mm）采集；采用采水器收集 1000 mL 水样，现场加 15 mL 鲁哥试剂固定，放置过夜；用吸管利用虹吸原理导出上部水，底部收集到 100 mL 试样瓶中寄到检测站进行定性定量分析。浮游动物采用 13#浮游生物网（网格宽 0.112 mm）；采用采水器采集 5L 水样，收集于 50 mL 试样瓶中，加入 5%甲醇固定浮游生物，寄到检测站。所有浮游生物的定性定量分析工作均由南京地理与湖泊研究所完成。

5.3 水生环境中砷的生物富集与转化

5.3.1 浮游生物多样性与丰度分布

浮游植物作为生态系统的主要初级生产者,其数量、种群组成的变化会对水生生态系统产生显著的影响。另外,浮游生物中的砷可以通过食物链传递对处于更高营养级的生物构成威胁。本研究在矿区水体中鉴定出的浮游植物类别有蓝藻门、硅藻门、甲藻门、金藻门、隐藻门、裸藻门、绿藻门(见图5.1)。可以看出在区域6和区域7,浮游植物的生长情况远远优于其他区域,总丰度为 $1200×10^4$ cells/L。其中,区域1、区域2、区域3、区域4、区域5中浮游植物主要为硅藻门、蓝藻门和绿藻门,总丰度仅为 $20×10^4$ cells/L。另外,浮游动物的鉴定结果与浮游植物的鉴定结果类似,在区域6和区域7中,浮游生物的含量远远高于其他区域,但是浮游动物的总丰度小于 2 ind/L。

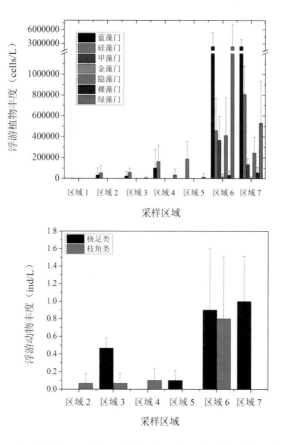

图 5.1 水体中的浮游生物（上：浮游植物；下：浮游动物）

The plankton in the freshwater from different areas

浮游生物是水生生态系统的重要组成部分。但是与其他研究结果相比，本研究区域水体中浮游生物的丰度远远低于其他报道。张楠（2013）研究发现，太湖不同区域水体中浮游植物的丰度为 $1057\times10^4 \sim 199398\times10^4$ cells/L。Yang 等人（2016）对我国两大淡水湖泊（太湖和巢湖）水体中浮游生物的研究发现，其水体中浮游植物的丰度为 $2300\times10^4 \sim 15000\times10^4$ cells/L。

浮游生物中砷的测量需要过滤，但是大量研究报道不同类型的滤膜（玻璃纤维滤膜、石英纤维滤膜、醋酸/硝酸纤维滤膜、聚四氟乙烯滤膜等）均含有一定量的砷（Helsen，2005；耿頔 等，2015），再加上实验条件有限，因此本研究中浮游生物中的砷及其砷形态并未做进一步分析。

5.3.2 水生生物对砷的富集特征及其影响因素

根据第 3 章，不同区域水体中砷含量差异较大，区域 3、区域 4、区域 5 水体中的砷含量显著高于区域 2 和区域 7。在本研究中，不同区域水生生物中的砷含量如图 5.2 所示。区域 3、区域 4 和区域 5 的水生生物中

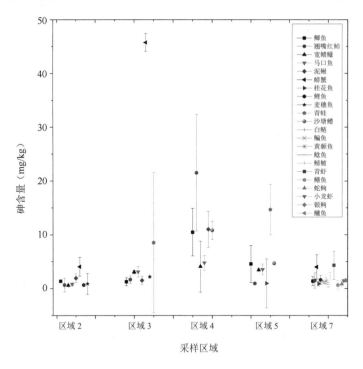

图 5.2 不同区域水生生物中的砷含量

The total arsenic concentration of aquatic organisms from different areas

砷含量相当高（变化范围分别为 1.17~45.75 mg/kg、4.07~21.53 mg/kg、0.94~14.66 mg/kg），远远高于区域 2 和区域 7 中水生生物中的砷含量（变化范围分别为 0.60~3.30 mg/kg、0.63~3.99 mg/kg）。在本研究中，水生生物对砷的富集受到多种因素的影响。

5.3.2.1　水生生物生存环境对砷富集的影响

目前关于淡水鱼类中砷含量的研究相对较多。在本研究中，区域 2 和区域 7 中淡水鱼类体内砷含量与大多数非污染区的淡水鱼类体内砷含量基本一致。Alamdar 等人（2017）研究发现，巴基斯坦 Chenab 河淡水鱼类体内砷含量范围为 0.23~1.21 mg/kg（湿重）。同样，Ciardullo 等人（2010）发现，意大利 Tiber 河淡水鱼类体内砷含量范围为 0.35~1.80 mg/kg。Ikemoto 等人（2008）则测定越南 Mekong Delta 淡水鱼类体内砷含量，结果表明砷含量范围为 0.32~2.3mg/kg。de Rosemond 等人（2008）测定加拿大 Back Bay 淡水鱼类体内砷含量范围为 0.57~1.15 mg/kg。Fu 等人（2011）对我国湖南锡矿山淡水鱼类体内砷含量进行分析，其范围为 0.28~1.32 mg/kg。Yang 等人（2017）检测分析发现，太湖和滇池淡水鱼体内砷含量范围为 0.14~1.88 mg/kg。Schaeffer 等人（2006）也报道，多瑙河淡水鱼体内砷含量范围为 0.25~1.37 mg/kg。Šlejkovec 等人（2004）检测出斯洛文尼亚淡水鱼体内砷含量范围为 0.08~1.23 mg/kg；Miyashita 等人（2009）也发现，日本 Hayakawa River 淡水鱼体内砷含量范围为 0.3~2.1 mg/kg。Cott 等人（2016）对加拿大淡水鱼体内砷含量进行分析，其范围为 0.1~2.8 mg/kg。Juncos 等人（2016）研究发现，阿根廷 Nahuel Huapi 湖泊淡水鱼类体内砷含量范围为 0.2~1.9 mg/kg；Hong 等人（2014）报道，韩国浦项市淡水鱼类体内砷含量范围为 0.64~4.1mg/kg；Phan 等人（2013）对泰国鲶鱼和黑鱼体内砷含量进行研究，发现其砷含量范围为 0.07~3.99 mg/kg。Liu 等人（2018）的研究也表明，中国黄河河口鱼类体内砷含量范围为 0.03~1.4 mg/kg。

而污染较为严重的区域3、区域4和区域5淡水鱼类的砷含量远远高于区域2和区域7。不同区域的同种淡水鱼类的砷含量也不同,如在严重污染的区域4中,翘嘴红鲌、泥鳅、鲫鱼的砷含量分别为21.52 mg/kg、11.02 mg/kg、10.48 mg/kg,远远高于其他区域中同种鱼类的砷含量。同样,在砷矿区上游污染较低的区域2,马口鱼和宽鳍鱲的砷含量分别为0.81 mg/kg和0.60 mg/kg,远远低于水污染较严重的区域3、区域4和区域5。

综上所述,同种水生生物在高砷污染区的水环境中,其对砷的富集也有一定程度的增加。该研究结果与前人的研究结果一致。Arain等人(2008)和Shah等人(2009)研究发现,在巴基斯坦污染河流Manchar Lake(水体砷含量为60~101 μg/L)中,淡水鱼类体内的砷含量范围为1.92~14.1 mg/kg。

对比砷污染相对较轻的区域2和区域7,以及污染较为严重的区域3、区域4、区域5,水体砷含量直接影响水生鱼类对砷的富集。这与前人的研究结果相同。Culioli等人(2009)发现,当水体砷含量分别为2330±835 μg/L、108±42 μg/L、43±27 μg/L、2.13±1.01 μg/L时,*Salmo trutta*肌肉组织中的砷含量也呈现下降趋势,分别为1.45±0.51 mg/kg、0.49±0.19 mg/kg、0.13±0.09 mg/kg、0.01±0.01 mg/kg。Jankong等人(2007)报道,泰国参照区(水体砷含量:1.4 μg/L)中淡水鱼*Channa striata*的砷含量为1.9±1.4 mg/kg,远远低于污染区(水体砷含量分别为550 μg/L和990 μg/L)中鱼类的砷含量(分别为13.1±1.0 mg/kg和22.2±2.2 mg/kg)。室内模拟研究实验也表明,鱼类中的砷含量水平随着添加砷含量的增加而增加(Suhendrayatna et al., 2001a, 2002; Kim and Kang, 2015)。

相对淡水鱼类,其他淡水水生生物的报道相对较少。在本研究中,区域7中小龙虾的砷含量为1.44 mg/kg,青虾的砷含量为2.80 mg/kg。该结果远远高于比利时市场上的小龙虾的砷含量(湿重0.12~0.17 mg/kg)(Ruttens et al., 2012)。Revenga等人(2012)也发现,安第斯山脉Moreno

湖小龙虾（*Samastacus* sp.）中的砷含量为0.35~0.87 mg/kg。但是，Juncos等人（2016）对阿根廷Nahuel Huapi湖泊小龙虾（*Samastacus spinifrons*）中的砷含量进行测定，发现其体内砷含量为0.4~6.4 mg/kg。Gedik等人（2017）的研究表明，美国路易斯安那小龙虾（*Procambarus clarkii*）肌肉组织中的砷含量为0.52~3.34 mg/kg（土壤砷含量为16.84~19.71 mg/kg）。Cui等人（2011）分析中国黄河河口虾（*Fenneropenaeus chinensis*）中的砷含量，发现达到7.5 mg/kg。Hong等人（2014）分析发现，韩国浦项市小虾（*Neocaridina denticulata*）中的砷含量为5.1~5.6 mg/kg。Williams等人（2009）研究发现，澳大利亚维多利亚小龙虾（*Cherax destructor*）中的砷含量随着环境中砷含量的增加而增加，当水体砷含量为300 μg/L、沉积物砷含量为190 mg/kg时，其体内砷含量最高可达449 mg/kg。在本研究中，在区域3中螃蟹的砷含量为45.75 mg/kg，高于区域2和区域7（3.30~3.99 mg/kg），也高于Fu等人（2011）报道的中国湖南锡矿山螃蟹中的砷含量（1.30~4.60 mg/kg）。Cui等人（2011）研究发现，中国黄河河口螃蟹（*Eriocheir sinensis*和*Mactra veneriformis*）中的砷含量为4.33~4.77 mg/kg。Juncos等人（2016）发现，阿根廷Nahuel Huapi湖泊螃蟹（*Aegla* sp.）中的砷含量范围较大，为1.5~12.4 mg/kg。而Revenga等人（2012）发现，安第斯山脉Moreno湖螃蟹（*Aegla* sp.）中的砷含量为2.75~3.97 mg/kg。在本研究中，青蛙中的砷含量分别为：区域3为8.54 mg/kg，区域5为14.66 mg/kg，远远高于Fu等人（2011）报道的中国湖南锡矿山青蛙中的砷含量（0.22~0.52 mg/kg），也高于Schaeffer等人（2006）报道的匈牙利多瑙河青蛙（*Rana* sp.）中的砷含量（2.52 mg/kg）。综上所述，水生生物对砷的富集与环境砷污染程度密切相关。

5.3.2.2　水生生物种间差异对砷生物富集的影响

同一区域不同种类的水生生物肌肉组织中砷的富集量也存在明显的差异。这与大多数研究结果一致（Ikemoto et al.，2008；Juncos et al.，2016；

Perera et al., 2016）。在砷污染程度较低的区域 2 和区域 7，螃蟹中砷含量最高，为 3.30～3.99 mg/kg。在砷污染严重的区域 3、区域 4、区域 5，螃蟹中砷含量最高（45.75 mg/kg）。Fu 等人（2011）对中国湖南锡矿山多种生物中的砷含量测定发现，螃蟹中的砷含量（1.30～4.60 mg/kg）高于鱼类及青蛙等水生动物（<1 mg/kg）。在本研究中，不同区域均表现为沙塘鳢、泥鳅、青虾、螃蟹等底栖生物中的砷含量较高。大量研究都表明，底栖无脊椎生物中的砷含量高于鱼类。Culioli 等人（2009）发现，Presa 河底栖生物中的砷含量最低为 6.30 mg/kg，最高可达 1928 mg/kg，远远高于淡水鱼类 *Salmo trutta* 中的砷含量（0.88～2.93 mg/kg）。Yang 等人（2017）发现太湖和滇池底栖生物中的砷含量为 4.33～12.32 mg/kg，远远高于淡水鱼类（<1 mg/kg）。Fliedner 等人（2014）测定发现，德国河流底栖生物贻贝 *Dreissena polymorpha* 中的砷含量为 2.25～12.78 mg/kg，远远高于鲂类 *Abramis brama* 中的砷含量（0.1～0.97 mg/kg）。Juncos 等人（2016）则发现，阿根廷 Nahuel Huapi 湖泊底栖生物中的砷含量为 0.4～67.6 mg/kg，远高于淡水鱼类（0.4～5.88 mg/kg）。Schaeffer 等人（2006）发现，多瑙河底栖生物贻贝 *Unio pictorum* 中的砷含量为 9.31～11.6 mg/kg，远远高于淡水鱼类中的砷含量（< 1.37 mg/kg）。同样，Roig 等人（2013）发现，Francolí 河流沉积物中的砷含量为 10～16 mg/kg，底栖生物中的砷含量最高可达 6.02 mg/kg，除底栖生物外，不同种类的淡水鱼中的砷含量也不同，其中桂花鱼中的砷含量最低。Has-Schön 等人（2015）对 Buško Blato 水库中鱼类的砷含量研究表明，鲶鱼 *Sylurus glanis* 中的砷含量（0.413±0.05 mg/kg）远远高于鲤鱼类 *Cypinus carpio*（0.224±0.03 mg/kg）。可见，不同种类的水生生物对砷的富集量存在明显的差异。

5.3.2.3 水生生物组织差异性对砷生物富集的影响

很多研究都表明，鱼类不同组织中的砷含量存在显著差异，主要表现为肝脏>肌肉（Maher et al., 1999; Suhendrayatna et al., 2001b; Schaeffer et al., 2006; Jankong et al., 2007; Arain et al., 2008; de Rosemond et al., 2008; Shah et al., 2009a; Fu et al., 2010; Revenga et al., 2012; Kim and Kang, 2015; Has-Schön et al., 2015; Cott et al., 2016; Dovick et al., 2016; Perera et al., 2016; Juncos et al., 2016; Yang et al., 2017）。在本研究中，不同区域水生生物不同组织中的砷含量如图 5.3 所示。在区域 7 采集的桂花鱼不同组织中，砷含量由高到低依次为鱼肝>鱼鳃>肌肉>鱼卵。Gedik 等人（2017）报道了美国路易斯安那小龙虾（*Procambarus clarkii*）不同组织中的砷含量，其外壳中的砷含量高于肌肉中的砷含量。但是，区域 3 中螃蟹肌肉中的砷含量远远高于外壳中的砷含量。对于大多数水生生物，我们采集了鱼皮和肌肉组织，从图 5.3 中可以看出，鱼皮中的砷含量要远高于肌肉中的砷含量。但是到目前为止，关于鱼类可食用部分——鱼皮的研究相对较少，研究结果也相反。Fu 等人（2010）发现，中国湖南锡矿山淡水鱼类不同组织中的砷含量分别为肝脏（0.25±0.03 mg/kg）>肌肉（0.16±0.01 mg/kg）>鱼皮（0.14±0.03 mg/kg）。Yang 等人（2017）发现我国太湖淡水鱼类不同组织中的砷含量分别为肝脏（5.47±3.27 mg/kg）>肌肉（0.91±0.52 mg/kg）>鱼皮（0.62±0.37 mg/kg）。

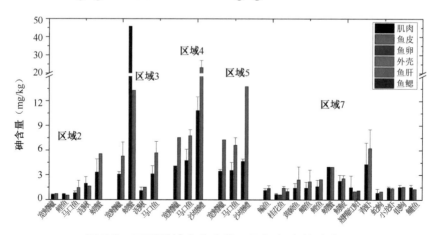

图 5.3 不同区域水生生物不同组织中的砷含量

The total arsenic concentration in the tissues of aquatic organisms

5.3.2.4 水生生物体重、体长对砷生物富集的影响

本研究选取数量较多的水生生物,分别对其组织中的砷含量与体长、体重的关系进行了分析(见图 5.4)。结果显示,宽鳍鱲、翘嘴红鲌、泥鳅、

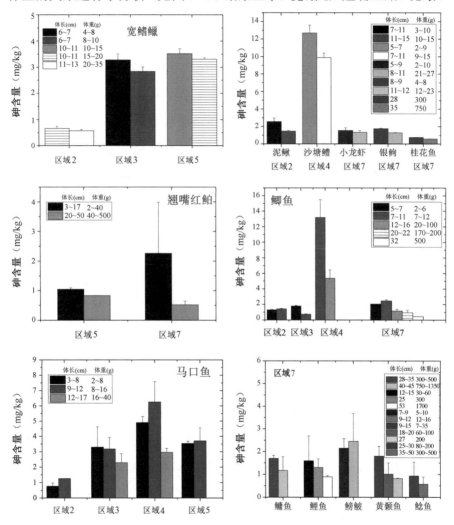

图 5.4 水生生物中的砷含量与体长、体重的关系

Relationship between total arsenic concentration and body size (length and weight) of aquatic organisms

沙塘鳢、银鮈、桂花鱼、鲫鱼、小龙虾、鳙鱼、鲶鱼、黄颡鱼、鲤鱼基本上随着体重、体长的增加，其肌肉中的砷含量有一定的下降，与大多数研究结果一致。Cott 等人（2016）对加拿大鲑鱼中的砷含量研究发现，鲑鱼中的砷含量随着体长的增加而降低。Liu 等人（2018）对中国黄河河口鱼类中的砷含量研究发现，砷含量与体重、体长均呈现明显的负相关关系（体重：R^2=0.182，$p<0.01$；体长：R^2=0.071，$p<0.05$）。Watanabe 等人（2008）报道日本 Ginzan 河水生生物中的干重与砷含量呈现明显的负相关关系（$p<0.01$）。但 Gedik 等人（2017）对美国路易斯安那小龙虾（*Procambarus clarkii*）中的砷含量研究发现，肌肉组织中的砷含量与体重呈现明显正相关关系。在本研究中，部分地区的马口鱼中砷含量与体重、体长的关系并不明显。

5.3.2.5 水生生物食性、栖息水层对砷生物富集的影响

根据不同区域水体中的砷含量，本研究将区域 2 和区域 7 分为一类，将区域 3、区域 4、区域 5 分为一类。水生生物中的砷含量与其食性、栖息水层的关系如图 5-6 所示。区域 3、区域 4、区域 5 采集的水生生物中主要为杂食性生物和肉食性生物，就砷含量而言，肉食性>杂食性。但是，在区域 2 和区域 7 中，就砷含量而言，滤食性>草食性>肉食性。就栖息水层而言，水生生物中的砷含量表现出一致的变化规律，在不同区域总体的变化趋势为：底栖>中层>上层。这主要是因为沉积物是水体中重金属的储存库，底层水生生物长期的体表砷暴露造成体内砷含量提高。另外，底层水生生物的食物来源大多为砷含量较高的碎屑、软体动物等，通过食物链的传递，最终砷富集在底栖生物体内。

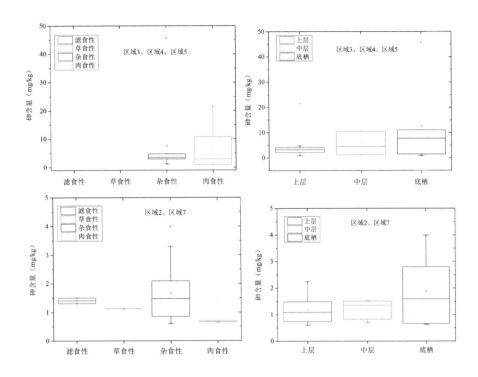

图 5.5 水生生物中的砷含量与食性、栖息水层的关系

Relationships between total arsenic concentration and feeding habit, habitat type of aquatic organisms

5.3.3 不同营养级水生生物对砷的富集效应

5.3.3.1 水生生物 $\delta^{13}C$ 值分布特征

水生生物肌肉组织中碳氮同位素分布如图 5.6 所示。$\delta^{13}C$ 值变化范围为 -30.59‰~-15.07‰，与文献报道的基本一致。王玉玉等人（2009）研究发现，鄱阳湖节肢动物和鱼类中 $\delta^{13}C$ 值范围为 -31.3‰~-23.4‰。李斌

等人（2009）测得小江下游无脊椎动物和鱼类中$\delta^{13}C$值范围为-27.7‰～-18.4‰，且呈现一定的季节性变化。Ikemoto等人（2008）发现，越南Mekong Delta水生生物中$\delta^{13}C$值范围为-29‰～-24‰。Wang等人（2012）报道，太湖水生生物中$\delta^{13}C$值范围为-29‰～-16‰。Saigo等人（2015）和Marchese等人（2014）对Paraná河水生生物中$\delta^{13}C$值分析发现，其范围为-28.5‰～-23.1‰。Pereira等人（2010）发现，巴西南部Mãe-Bá泄湖水生生物中$\delta^{13}C$值范围为-33.9‰～-20.2‰。Juncos等人（2016）报道，阿根廷Nahuel Huapi湖泊淡水鱼类的$\delta^{13}C$值范围为-31.2‰～-15.6‰。Liu等人（2018）分析发现，中国黄河河口鱼类中的$\delta^{13}C$值范围为-27.6‰～-16.2‰。同样，Cui等人（2011）也对中国黄河河口水生生物中的$\delta^{13}C$值进行测定，发现其范围为-25.34‰～-19.98‰。

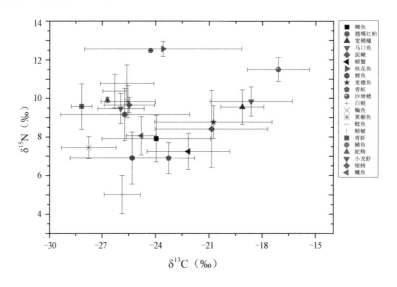

图5.6 水生生物肌肉组织中碳氮同位素分布

Carbon and nitrogen stable isotope of aquatic organisms

同种生物的$\delta^{13}C$值相差甚远，如鲫鱼、翘嘴红鲌、泥鳅、桂花鱼、鲤鱼和麦穗鱼等，其$\delta^{13}C$值的变化范围可达6‰以上。$\delta^{13}C$值是反映生物的食物来源的重要指标,说明不同区域的同种水生生物其食物来源差异明显,这也与水生生物的食物多样性有关。

我们对不同水域同种生物的δ^{13}C值进行分析,发现区域7中桂花鱼、鲫鱼、鲤鱼、螃蟹、翘嘴红鲌的δ^{13}C值远低于其他区域。这主要由于区域7中,δ^{13}C值低的浮游生物相较于其他地区更为丰富。在各采样区域,宽鳍鱲、马口鱼、泥鳅、麦穗鱼、沙塘鳢的δ^{13}C值较高(-20.86‰~-17.10‰),青虾、鳑鲏、鳊鱼、蛇鉤的δ^{13}C值较低(-28.16‰~-26.28‰)。这与不同水生生物食物来源差异有关,滤食性生物主要摄取微囊藻类、浮游动物等δ^{13}C值较低的饵料,而底栖生物食物来源主要为外来碎屑、软体动物等δ^{13}C值高的饵料。

5.3.3.2 水生生物δ^{15}N值分布特征

不同区域的水生生物肌肉组织中的δ^{15}N值变化范围为4.31‰~12.98‰(见图5.6),与文献报道的基本一致。王玉玉等人(2009)发现,鄱阳湖节肢动物和鱼类中δ^{15}N值范围为6.1‰~15.4‰。李斌等人(2009)发现,小江下游无脊椎动物和鱼类中δ^{15}N值范围为7.1‰~13.65‰。Ikemoto等人(2008)报道,越南Mekong Delta水生生物中δ^{15}N值范围为9‰~18‰。Wang等人(2012)测得太湖水生生物中δ^{15}N值范围为10‰~18‰。Saigo等人(2015)和Marchese等人(2014)研究发现,Paraná河水生生物中δ^{15}N值范围为6.50‰~11.97‰。Pereira等人(2010)发现,巴西南部Mãe-Bá泄湖水生生物中δ^{15}N值范围为7.3‰~11.1‰。Liu等人(2018)和Cui等人(2011)研究发现,中国黄河河口水生生物中δ^{15}N值范围为5.1‰~14‰。Juncos等人(2016)发现,阿根廷Nahuel Huapi湖泊淡水鱼类中δ^{15}N值范围为1.8‰~15.6‰。

其中,区域2的水生生物中的δ^{15}N值范围为5.96‰~11.22‰,区域3的水生生物中的δ^{15}N值范围为5.68‰~10.82‰,区域4的水生生物中的

$\delta^{15}N$ 值范围为 6.99‰~12.77‰，区域 5 的水生生物中的$\delta^{15}N$ 值范围为 6.33‰~12.82‰，区域 7 的水生生物中的$\delta^{15}N$ 值范围为 4.31‰~12.98‰。不同区域同种生物的$\delta^{15}N$ 值变化范围一般<2‰。但是，泥鳅的$\delta^{15}N$ 值变化范围较大，在不同区域可达到一个营养级，说明不同生境下泥鳅在食物链中的营养位置发生了改变。同时，不同区域的生物种群的组成不同，这可能与环境中砷污染有关，但仍需要进一步的研究。$\delta^{15}N$ 值会随着营养级的升高而出现稳定富集现象，不同营养级间的$\delta^{15}N$ 富集值平均为 3.4‰（Vander Zanden et al., 1997, 1999; Post, 2002b）。以此推算，在区域 2、区域 3、区域 4、区域 5、区域 7 采集的水生生物的营养级跨度分别为 1.55、1.51、1.70、1.91、2.55。

一般认为，肉食性生物处于食物链顶端，杂食性生物次之，滤食性生物位于食物链的底端。在本研究中，水生生物中氮同位素含量与栖息水层、食性的关系如图 5.7 所示。肉食性水生生物的氮同位素含量最高，杂食性水生生物的氮同位素含量次之，滤食性水生生物的氮同位素含量最低。这与 Wang 等人（2012）的研究结果一致，太湖不同种水生生物中的$\delta^{15}N$ 值排序：大型水生植物 < 浮游生物 < 底栖生物 < 草食性鱼类 < 杂食性鱼类 < 肉食性鱼类。但是，水生生物的栖息水层与氮同位素含量并没有明显的关系。

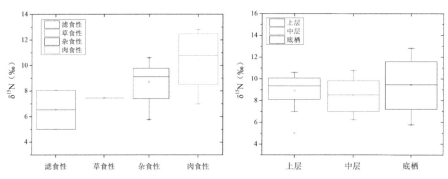

图 5.7 水生生物中氮同位素含量与栖息水层、食性的关系

The relationships between nitrogen stable isotope and food habits, habitat type of aquatic organisms

5.3.3.3 砷沿着水体食物链的生物放大效应

水生生物中的砷含量与$\delta^{15}N$的关系如图5.8所示。所得回归方程为,区域2:$y=-0.20x+2.87$($R^2=-0.17$,$p>0.05$);区域3:$y=-4.53x+45.15$($R^2=0.17$,$p>0.05$);区域4:$y=-2.69x+36.72$($R^2=0.39$,$p>0.05$);区域5:$y=-1.30x+16.83$($R^2=0.29$,$p>0.05$);区域7:$y=-0.10x+2.50$($R^2=-0.02$,$p>0.05$)。由此可见,水生生物中的砷含量(y)与$\delta^{15}N$(x)存在一定的负相关关系,但是并不显著。

前人的研究也发现,随着营养级的增加,砷含量有一定的下降趋势。Revenga等人(2012)对安第斯山脉Moreno湖水生生物中砷含量做对数转换后,发现与$\delta^{15}N$呈明显的负相关关系(斜率:-0.23,$R^2=0.31$,$p<0.05$)。Vizzini等人(2013)对地中海Stagnone di Marsala水生生物中砷含量做对数转换后,与$\delta^{15}N$呈明显的负相关关系(斜率:-0.31,$R=-0.74$,$p<0.01$)。Pereira等人(2010)对巴西南部Mãe-Bá泄湖水生生物中砷含量做对数转换后,与$\delta^{15}N$呈明显的负相关关系(斜率:-0.36,$R^2=0.54$,$p<0.01$)。Juncos等人(2016)也发现,对阿根廷Nahuel Huapi湖泊水生生物中砷含量做对数转换后,与$\delta^{15}N$呈明显的负相关关系($p<0.01$)。Cui等人(2011)研究发现,对中国黄河河口水生生物中砷含量做对数转换后,与$\delta^{15}N$呈一定的负相关关系,但并不显著(斜率:-0.39,$R=0.43$,$p=0.11$)。Watanabe等人(2008)发现,日本Ginzan河水生生物中砷含量与$\delta^{15}N$呈一定的负相关关系,但并不显著。

但是,Zhang和Wang等人(2012)对中国沿海6个城市鱼类中的砷含量与$\delta^{15}N$进行研究,发现两者之间并没有明显的相关性,与本文研究结果相反。Ikemoto等人(2008)对越南Mekong Delta淡水鱼类中砷含量做对数转换后,与$\delta^{15}N$呈一定的正相关关系(斜率:0.064,$R^2=0.36$,$p>0.05$)。Liu等人(2018)的研究也表明,对中国黄河河口鱼类中砷含量做对数转换后,与$\delta^{15}N$呈一定的正相关关系($R^2=0.009$,$p>0.05$)。因此,水生生物中砷是否具有"稀释效应",还需要更多的研究。

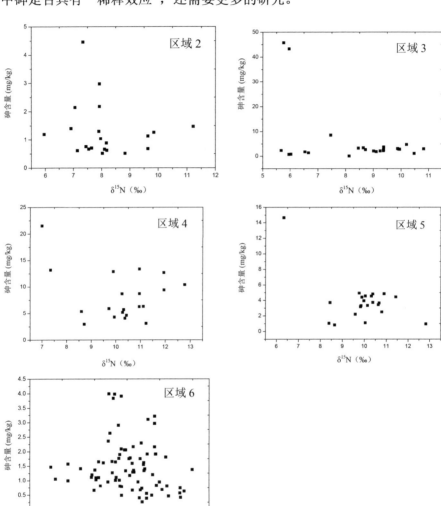

图 5.8 水生生物中的砷含量与 $\delta^{15}N$ 的关系

The relationship between arsenic concentration and nitrogen stable isotope

根据公式：生物富集系数（BF）=生物体中的砷含量（mg/kg，干重）/水体中的砷含量（mg/L），区域 3、区域 4、区域 5 水体中砷含量无明显差异，采集的水生生物对砷的富集系数范围大多为 0.46~8.49；而区域 2 和区域 7 水体中砷含量较低，水生生物对砷的富集系数范围大多为 30.09~305。Culioli 等人（2009）研究发现，Bravona 河水生生物中，鱼类 *Salmo trutta* 的砷富集系数为 0.04~10.5，而底栖生物的砷富集系数为 23~557。Fu 等人（2011）发现，中国湖南锡矿山鱼类的砷富集系数为 1.84~193；Zhang 等人（2013）报道，中国太湖和滇池鱼类的砷富集系数分别为 56~152、19~69；Hong 等人（2014）发现，韩国浦项市采样点水体中的砷含量为 0.69~0.90 μg/L，淡水鱼类的砷富集系数为 711~5466。众多研究表明，水生生物对砷的富集能力与水体砷污染程度密切相关。

5.3.4　水生生物中的砷形态及其转化

5.3.4.1　水生生物中的砷形态特征

采用 HPLC-ICP-MS 对采集的水生生物进行砷形态分析，部分代表样品的质谱图如图 5.9 所示。其中，未知砷 1 在水生生物中大量存在，其流出时间比 AsB 长，但是在 As(III)前流出，与实验室使用的标准物质的流出时间均不相符；未知砷 2 的流出时间与紫菜中的主要砷形态流出时间一致（图谱未给出），推断为砷糖。

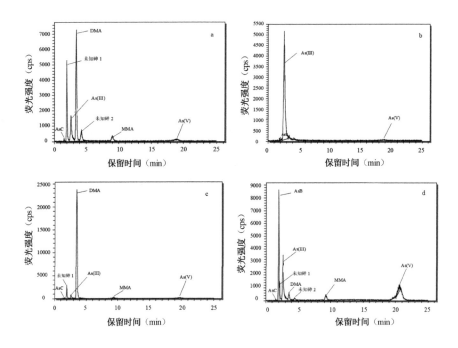

图 5.9 水生生物的 HPLC-ICP-MS 质谱图

(a: 马口鱼—区域 5; b: 青虾—区域 7; c: 沙塘鳢—区域 4; d: 鲫鱼—区域 4)

HPLC chromatograms of the methanol-water-extractable As compounds in aquatic organisms

水生生物中的砷形态如表 5.2 所示，主要包括未知砷 1、DMA，还含有少量的 AsC、MMA、AsB 和未知砷 2，无机砷 As(III)和 As(V)的比例相对较低。水生生物不同组织间的砷形态也不一样，如小龙虾肌肉中的砷形态主要为有机砷，而外壳中主要为 As(III)（约 68%）。但是在本研究中，鱼肉和鱼皮中的砷形态基本一致。

表 5.2 HPLC-ICP-MS 方法分析的水生生物中的砷形态

HPLC-ICP-MS data: The extracted arsenic species of aquatic organisms

样品名称	采样区域	部位	AsC	AsB	未知砷1	提取砷形态比例（%） As(III)	DMA	未知砷2	MMA	As(V)	提取率(%)	柱回收率(%)
白鲢	7	肌肉	10.56±18.29		59.38±15.98	1.39±2.41	18.85±7.40	3.67±6.36	tr	6.15±10.64	31.92±17.64	76.17±19.09
鳊鱼	7	鱼卵	0.19±0.46		75.87±22.49	9.13±6.17	6.04±4.20	2.60±6.38		6.17±10.23	49.50±18.07	75.28±12.12
	7	肌肉			67.45±11.65	3.64±5.09	20.24±5.43	1.20±2.94		7.47±9.45	55.71±8.33	64.37±23.67
	5	肌肉	16.31	1.98	23.18		29.49	tr	29.04		52.23	48.99
桂花鱼	7	肌肉	19.46±12.31	14.08±19.91	1.78±2.52	13.41±18.96	25.07±26.32	20.19±28.55	6.01±8.49		27.84±24.92	60.70±12.93
	7	鱼肝	9.43		30.94	16.08	17.92		25.63		32.78	66.75
	7	鱼卵	2.04		15.29	27.46	2.04	2.04		51.11	45.44	54.86
	7	鱼鳔	tr		18.13±16.80	13.66±2.92	20.15±5.51	2.31±3.27	tr	45.75±21.96	38.50±7.74	69.34±4.58
黄颡鱼	7	肌肉	3.24±4.06	2.77±7.34	83.58±9.74	6.14±5.17	1.41±1.77		0.49±1.30	2.37±4.26	32.57±27.37	91.81±19.68
	7	鱼皮	1.41±1.27		80.94±21.91	4.14±7.17	13.51±23.39				41.94±11.17	73.91±25.93
	2	肌肉			62.68			37.32			64.45	77.09
鲫鱼	3	肌肉			73.11±14.40	tr	3.42±4.84	23.47±9.56			45.40±31.02	92.32±28.77
	4	肌肉	0.44±0.28		39.79±11.51	13.30±4.62	8.07±8.12	1.45±0.33	4.65±0.94	32.30±14.67	47.44±22.40	98.92±16.73
	5	肌肉	1.70		38.88	15.30	18.96	5.12	3.32	16.72	47.16	70.07
	7	肌肉	3.26±2.40	1.39±3.58	47.10±29.09		16.26±19.84	28.57±31.49	tr	3.42±8.19	35.60±18.12	68.26±17.63
	7	鱼皮	1.86		32.05		12.01	54.08			42.70	86.39

续表

样品名称	采样区域	部位	AsC	AsB	提取砷形态比例（%） 未知砷1	As(III)	DMA	未知砷2	MMA	As(V)	提取率（%）	柱回收率（%）
宽鳍鱲	2	肌肉	tr	1.50±2.59	36.61±1.07	tr	11.93±8.59	48.79±9.51		1.17±2.04	19.40±5.45	68.72±4.20
	2	鱼皮			64.97		3.68	31.35			53.70	54.75
	3	肌肉	0.29±0.23		37.34±6.00	1.84±0.13	53.07±4.66	4.01±1.20	3.45±0.49	tr	22.24±1.53	114.30±2.45
	3	鱼皮	tr		64.01±9.61	5.37±1.71	26.67±5.73	2.39±3.38	1.56±2.21		50.52±9.87	94.77±4.45
	4	肌肉	0.46		25.46	4.75	56.54	9.77	3.03		19.00	87.55
	4	鱼皮	0.19±0.27		60.04±5.35	6.73±1.51	23.35±5.13	2.92±4.13	6.77±2.66	tr	49.31±17.01	97.68±19.94
	5	肌肉	0.26±0.36	2.12±3.00	36.05±3.51	5.89±0.96	40.47±3.35	4.10±0.96	6.75±2.16	4.36±1.04	21.10±4.65	95.95±14.34
	5	鱼皮	5.13	9.93	54.95		14.12	15.87			47.63	94.78
鲤鱼	2	肌肉		5.40	94.60		tr	tr			34.39	76.35
	2	鱼皮	3.08±3.11	0.74±1.49	48.93±30.02	3.09±4.22	33.26±33.29	3.65±4.28	1.37±2.75	5.88±7.50	40.99±26.47	81.35±28.45
	7	肌肉	tr		49.95	6.43	5.35	tr		38.27	20.71	80.86
	7	鱼皮	0.36±0.58		27.60±5.72	2.56±4.56	18.51±16.39	46.51±18.26	1.06±2.28	3.40±0.80	46.27±19.95	91.77±23.22
马口鱼	2	肌肉	0.30±0.52		46.79±8.75	tr	10.54±8.98	33.25±4.43	tr	9.12±15.80	63.85±15.41	105.79±12.91
	3	肌肉	0.36±0.36		44.41±8.74	3.43±3.43	41.14±10.09	3.28±3.30	3.81±2.60	3.57±4.58	29.03±20.89	103.51±7.34
	3	鱼皮	0.13±0.23		69.04±11.51	5.28±4.63	19.20±2.92	tr	2.84±3.47	3.51±6.07	53.98±10.82	89.81±16.07
	4	肌肉	0.98±1.35		39.60±3.78	5.98±2.57	41.04±5.94	3.39±0.94	5.84±2.54	3.17±2.91	32.19±24.74	104.51±3.86
	4	鱼皮	0.32		39.03	23.57	19.99	2.53	7.56	7.00	27.62	112.15
	5	肌肉	3.64±9.46	2.95±5.10	39.49±6.73	6.21±2.17	33.06±12.23	10.04±5.72	2.37±1.99	2.24±4.00	29.89±16.61	93.73±16.91
	5	鱼皮	0.56±0.59		65.37±4.45	9.08±1.55	21.35±2.88	1.43±2.48	1.16±2.00	1.05±1.83	42.30±4.71	81.78±15.45

第5章 水生环境中砷的生物富集与转化

续表

样品名称	采样区域	部位	AsC	AsB	未知砷1	As(III)	提取砷形态比例 (%) DMA	未知砷2	MMA	As(V)	提取率(%)	柱回收率(%)
麦穗鱼	2	肌肉	3.31		38.86	22.32	4.95	30.55	tr	tr	52.19	68.95
	3	肌肉			54.42	10.94	11.93	14.14	8.58	tr	44.93	94.32
泥鳅	2	肌肉	0.22±0.38	5.80±2.38	24.23±1.21	10.41±9.03	48.61±22.04	4.30±5.81	1.41±2.45	5.02±5.30	29.93±16.15	101.95±18.30
	2	鱼皮	tr		39.23		60.77				46.64	89.39
	3	肌肉		1.05±1.49	25.81±21.38	1.08±1.53	60.13±21.77	7.77±9.28	2.67±3.76	1.49±2.11	65.76±10.70	98.01±10.92
	3	鱼皮			32.96		67.04				28.36	92.02
	4	鱼皮	0.19±0.27		27.15±3.67	15.15±6.09	21.59±18.56	0.72±1.02	15.77±10.56	19.43±0.48	30.54	78
	7	鱼卵		24.59		tr	75.41				52.87	85.26
鲶鱼	7	肌肉	4.05±6.66	73.43±30.75	9.89±17.63		10.02±9.89		2.93	2.61±6.41	35.67±18.12	63.92±26.54
螃蟹	2	肌肉	2.53	18.37	21.31	5.93	24.30	11.06		13.57	25.01	76.26
	3	肌肉	0.35		6.33	69.92	5.63		tr	17.76	72.05	80.10
鳑鲏	7	肌肉	8.94±4.64	29.78±6.98	3.74±6.48	31.95±5.34	23.47±6.87	2.12±3.65		8.43±11.91	29.93±3.25	54.33±19.24
	7	肌肉	1.10±1.47	3.75±5.21	56.78±12.49	6.80±7.80	22.16±20.51	0.98±2.18			38.67±2.71	76.94±7.63
	7	鱼皮			77.64		22.36			tr	54.42	53.18
翘嘴红鲌	2	肌肉	tr		71.30	3.01	23.50	2.18		tr	29.43	62.37
	3	肌肉			32.94	10.43	19.94	36.69		tr	32.86	84.80
	4	肌肉	0.40		26.00	28.70	10.21		13.07	21.61	41.10	72.44
	7	鱼皮	4.77		32.70	tr	31.96	30.56			48.20	73.71
	5	肌肉		6.32±8.28	50.30±29.10	tr	13.27±7.30	27.54±31.72	tr	2.57±3.61	67.69±10.28	56.95±7.07
	7	肌肉	2.93±5.08	16.15±27.97	40.71±13.83	4.72±5.08	18.25±12.09	7.21±9.23	2.38±2.24	7.65±7.50	54.76±26.76	96.77±8.63
	7	鱼卵			38.04	tr	9.96	30.41		21.58	14.10	71.34

续表

样品名称	采样区域	部位	AsC	AsB	未知砷1	提取砷形态比例（%） As(III)	DMA	未知砷2	MMA	As(V)	提取率（%）	柱回收率（%）
青蛙	3	肌肉	0.60	0.44	1.22	16.35	41.42		28.65	11.31	33.43	59.19
	5	肌肉	1.50	0.44	1.48	35.77	40.96		9.77	10.08	66.33	90.84
青虾	7	肌肉	6.62±9.36	12.08±17.08		81.30±26.44					40.72±9.07	96.87±30.30
沙塘鳢	4	肌肉	0.72±0.21	1.00±0.97	7.50±0.68	1.83±1.03	85.04±1.41		1.19±1.49	2.72±0.85	19.39±1.49	115.02±5.27
	4	鱼皮	0.76±0.31		39.11±8.47	4.55±1.68	47.05±4.41		1.48±2.10	7.05±0.60	53.87±8.19	94.53±0.57
鳍鱼	5	肌肉	1.15	21.77	64.52	2.62	25.38		0.19	6.14	64.88	108.83
	7	肌肉	2.48				75.75				46.44	95.30
蛇鮈	7	肌肉		9.18±12.98	78.13±17.35		12.69±4.37				41.17±15.04	53.41±9.10
	7	鱼皮			47.51		52.49				24.50	96.57
小龙虾	7	外壳	31.85±14.59	29.90±12.66			16.10±4.02			22.15±23.24	33.29±8.09	43.54±9.20
银鲴	7	肌肉	16.13±5.05	11.70±9.01	3.20±4.53	67.78±20.27	1.19±1.69			tr	46.01±27.87	34.68±16.59
	7	肌肉	5.06±6.65		53.12±10.06	13.31±3.33	9.79±7.65	1.85±3.21	9.88±2.54	6.99±6.53	27.83±2.09	72.45±15.30
	7	鱼皮	6.48±9.16		41.98±3.81	43.69±1.87	2.85±4.03			5.00±7.07	47.91±24.26	77.89±10.96
鳙鱼	7	肌肉	1.09±1.94	1.55±3.79	35.05±18.57	tr	19.05±11.26	35.69±19.32		7.57±8.47	35.28±14.97	71.36±25.66
	7	鱼皮			41.30		13.94	44.76		tr	36.73	82.91

不同鱼类中的砷形态差异显著。在本研究中，鲶鱼肌肉中的砷形态主要为 AsB（约 73%）。Šlejkovec 等人（2006）也发现，斯洛文尼亚鲶鱼（*Silurus glanis*）中只检测到 AsB。大量研究发现，淡水鱼类中的砷形态主要为 AsB。Ciardullo 等人（2010）对淡水鱼类鳗鱼（*Anguilla anguilla* L.）、鲻鱼（*Mugil cephalus* L.）、鲢鱼（*Leuciscus cephalus* L.）肌肉中的砷形态分析发现，其主要为 AsB（>95%）。Miyashita 等人（2009）报道，日本 Hayakawa 河草食性鱼类 *Plecoglossus altivelis*，以及杂食性鱼类 *Cobitis biwae*、*Rhinogobius* sp.、*Sicyopterus joaponicus* 中的砷形态主要为 AsB（>52%）。Ruttens 等人（2012）对比利时市场上的淡水鱼类中的砷形态分析发现，鲈鱼、梭鲈和鲑鱼中只检测到 AsB。Hong 等人（2014）在韩国浦项市淡水鱼类 *Misgurnus mizolepis*、*Rhinogobius giurinus* 中只检测到 AsB。Šlejkovec 等人（2006）对斯洛文尼亚鲑鱼（*Salmo marmoratus*、*Oncorhynchus mykiss*、*Salmo trutta m. fario*）和鲶鱼（*Silurus glanis*）中的砷形态分析发现，其主要为 AsB（>92%）。

区域 4 中沙塘鳢肌肉中的砷形态主要为 DMA（约 85%）。Yang 等人（2017）发现，中国太湖和滇池淡水鱼类中的砷形态主要为 DMA（>75%）。Jankong 等人（2007）发现，泰国肉食性鱼类 *Channa striata* 中大于 91%的砷形态为 DMA。Miyashita 等人（2009）分析发现，日本 Hayakawa 河杂食性鱼类 *Anguilla japonica* 中的砷形态主要为 DMA（77%）。Cott 等人（2016）对加拿大 Yellowknife 鲑鱼、鳕鱼肌肉中的砷形态分析发现，其砷形态主要为 DMA。Šlejkovec 等人（2006）对斯洛文尼亚鳕鱼（*Lota lota*）中的砷形态分析发现，其砷形态也主要为 DMA（75%）。Ruttens 等人（2012）对比利时市场上的淡水鱼类中的砷形态分析发现，鳗鱼中只检测到 MMA。

大多数研究表明，淡水鱼类中的砷形态非常复杂，并非单一砷形态占绝对优势，通常由多种砷形态组成。Šlejkovec 等人（2006）报道发现，斯洛文尼亚鲤鱼中的砷形态相对比较复杂，除 AsB、TMAO、DMA 外，还检测到大量未知砷。Jankong 等人（2007）也报道，泰国杂食性 *Danio regina*、*Rasbora heteromorpha* 和草食性 *Puntius orphoides* 中砷形态主要为 As(V)（34%~53%）、TMAO（20%~38%）和 DMA（18%~22%）。Ciardullo 等人（2010）研究发现鲤鱼（*Cyprinus carpio* L.）肌肉中的砷形态主要为 AsB（58%），还含有少量的砷糖（26%）。Miyashita 等人（2009）对日本 Hayakawa 河淡水杂食性鱼类 *Zacco platypus* 中的砷形态分析发现，其砷糖的含量达到 51%。de Rosemond 等（2008）发现，加拿大 Back Bay 的 5 种淡水鱼类 *Coregonus clupeaformis*、*Stizostedion vvitreum*、*Esox lucius*、*Catostomus commersoni* 和 *Catostomus catostomus* 肌肉组织中的提取砷中大于 73%的砷为 AsB（36%~73%）和 DMA（19%~57%），还含有少量的 As(III)和 As(V)。Hong 等人（2014）对韩国浦项市淡水鱼类中的砷形态分析发现，其主要由 As(III)、As(V)、AsB、DMA 等砷形态组成。

除淡水鱼类外，其他水生生物中的砷形态研究报道相对少见。在本研究中，青蛙中的砷形态主要为 DMA、As(III)、As(V)、MMA。这与 Schaeffer 等人（2006）的研究结果基本一致，该研究发现，多瑙河的青蛙中的砷形态主要为 As(III)（30%）和 TETRA（35%），还含有少量的 As(V)、MMA、DMA。在本研究中，小龙虾中的砷形态主要为 AsC、AsB、As(V)、DMA，而青虾中的砷形态主要为 As(III)。Williams 等人（2009）研究澳大利亚维多利亚小龙虾（*Cherax destructor*）的砷形态发现，不同砷污染程度的小龙虾肌肉中的砷形态不同，主要为 AsB、As(III)，还含有少量的 DMA、Tetra。Hong 等人（2014）研究发现，韩国浦项市小虾（*Neocaridina denticulata*）的砷形态主要为 AsB（84%）和 DMA（16%），但是另一部分小虾中的砷形态主要为 As(III)（77%），还含有少量的 AsB 和 DMA。Ruttens 等人（2012）发现，比利时市场上小龙虾中的砷主要为 AsB 和 DMA。

水生生物中总砷含量与提取砷形态比例之间的相关性分析如表 5.3 所示，可以发现，以下两两显著正相关：AsB 和 AsC（$r=0.375$，$p=0.011$），

As(III)和总砷（$r=0.626$，$p<0.0001$），As(V)和总砷（$r=0.527$，$p<0.0001$）。显著负相关的有：未知砷 1 和总砷（$r=-0.329$，$p=0.027$），未知砷 1 和 AsC（$r=-0.348$，$p=0.019$），未知砷 1 和 AsB（$r=-0.388$，$p=0.009$），未知砷 1 和 As(III)（$r=-0.443$，$p=0.002$），未知砷 1 和 DMA（$r=-0.380$，$p=0.010$），As(V)和 MMA（$r=-0.317$，$p=0.034$）。结果表明，水生生物中的砷含量越高，其体内无机砷 As(III)和 As(V)的比例也越高。在本研究中，砷含量相对较高的主要为底栖生物，如螃蟹中 As(III)比例可达约 69%，青虾中 As(III)比例可达约 81%，泥鳅中 As(V)比例可达约 19%，小龙虾中 As(V)比例可达约 22%。底栖生物基本上生活在底泥中，以浮游生物、腐殖质、泥沙等为食，而浮游生物中的砷形态主要为无机砷（Caumette et al.，2011，2012）。因此，底栖生物中无机砷比例相应也较高。另外，随着外界环境中砷含量的增加，水生生物中的砷含量相应增加，其无机砷形态比例也相应增加。例如，区域 3 螃蟹中 As(III)比例可达约 69%，远高于区域 2 和区域 7；区

表 5.3　水生生物中总砷含量与提取砷形态比例之间的相关性分析

Correlation between total arsenic concentration and extracted arsenic species in aquatic organisms

	总砷	AsC	AsB	未知砷 1	As(III)	DMA	MMA	未知砷 2	As(V)
总砷	1								
AsC	−0.169	1							
AsB	−0.169	0.375*	1						
未知砷 1	−0.329*	−0.348*	−0.388**	1					
As(III)	0.626**	0.001	−0.029	−0.443**	1				
DMA	−0.036	−0.115	−0.077	−0.380*	−0.254	1			
MMA	−0.288	−0.149	−0.185	0.142	−0.240	−0.294	1		
未知砷 2	0.167	0.109	−0.189	−0.274	0.065	0.138	−0.279	1	
As(V)	0.527**	0.103	−0.044	−0.202	0.225	−0.187	−0.317*	0.218	1

注：*表示在 0.05 水平上显著相关，**表示在 0.01 水平上显著相关。

域 3 中 As(Ⅴ)比例为 17%，也高于其他区域。区域 4 和区域 5 鲫鱼中 As(Ⅲ)和 As(Ⅴ)比例都高于区域 2 和区域 7。Suhendrayatna 等人（2001a）通过室内砷添加实验也发现，随着 As(Ⅲ)添加量的增加，虾（*Neocaridina denticulata*）和肉食性鱼类 *Tilapia mossambica* 中 As(Ⅲ)和 As(Ⅴ)的比例也相应增加。由此推测，水生生物对无机砷的转化能力随砷富集量的增加而降低。

5.3.4.2 XANES 砷形态分布特征

不同水生生物的提取率和柱回收率变化范围较大，提取率约从 14.10% 到 72%，柱回收率约从 34.68% 到 115%（见表 5.2），这说明水生生物中还有大量的砷未被提取，同时提取砷中有部分砷未被检测出或者未被鉴定出。因此，采用 XANES 对水生生物中的砷形态进行归一化处理，图谱如图 5.10 所示。

XANES 分析水生生物中的砷形态结果如表 5.4 所示。水生生物中的砷形态主要为 DMA、As(Ⅲ)-GSH、As(Ⅲ)和 As(Ⅴ)。不同水生生物中的砷形态也不一样。不同区域的马口鱼中的砷形态几乎一致，基本表现为以 DMA 为主要砷形态，还含有少量的 As(Ⅲ)和 As(Ⅲ)-GSH；区域 4 和区域 5 马口鱼采用 HPLC-ICP-MS 分析出的砷形态主要为未知砷 1、DMA，还含有少量的 As(Ⅲ)、MMA 和 As(Ⅴ)。XANES 分析的鲫鱼中的砷形态为 DMA、As(Ⅴ)和 As(Ⅲ)-GSH；HPLC-ICP-MS 的结果表现为未知砷 1、As(Ⅴ)、DMA、As(Ⅲ)。XANES 分析的泥鳅、翘嘴红鲌和青蛙中的砷形态基本一致，均表现为 DMA、As(Ⅴ)和 As(Ⅲ)-GSH；HPLC-ICP-MS 的结果表现为未知砷 1、DMA、As(Ⅲ)、MMA 和 As(Ⅴ)。XANES 分析的宽鳍鱲、鳑鲏、沙塘鳢和鲤鱼中的砷形态基本一致，均表现为 DMA、As(Ⅲ)和 As(Ⅲ)-GSH；HPLC-ICP-MS 的结果表现为未知砷 1、DMA，还含有少量的 As(Ⅲ)和 As(Ⅴ)。由于 As(Ⅲ)易与 S 配位结合，在本研究中，As(Ⅲ)-GSH 在所有样品中均被检测出，说明在水生生物中存在一定量的 As-S 配位的砷。但是，由于 HPLC-ICP-MS 方法有部分砷未被提取出来，还有部分提取砷未被检测出,因此,通过以上对比可知,HPLC-ICP-MS 分析的砷形态与 XANES 分析的砷形态结果基本一致。推测未被提取和未被检测出的砷形态可能为 As-S 配位的砷。

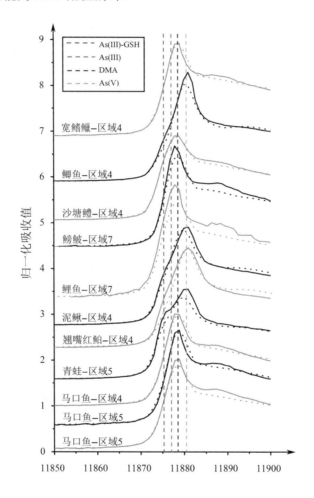

图 5.10　水生生物中砷形态的 XANES 归一化处理谱图

(实线表示原谱图,虚线表示拟合曲线)

XANES raw data (—) and fitting (---) for model As species and samples

表 5.4　XANES 分析水生生物中的砷形态结果

XANES fitting results for aquatic organisms

样品	采样点位	砷形态比例（%）				卡方检验
		As(III)-GSH	As(III)	As(V)	DMA	
鲫鱼	区域4	15.3	0	65.6	19.0	0.012
宽鳍鱲	区域4	13.2	15.8	0	71.0	0.002
沙塘鳢	区域4	30.5	13.7	0	55.8	0.001
鳑鲏	区域7	2.6	33.1	0	64.3	0.005
鲤鱼	区域7	26.3	18.4	0	55.4	0.067
泥鳅	区域4	21.6	0	36.2	42.2	0.011
翘嘴红鲌	区域4	19.4	0	44.6	36.0	0.017
青蛙	区域5	37.3	0	29.5	33.3	0.006
马口鱼	区域4	14.1	18.9	0	67.1	0.002
马口鱼	区域5	15.2	18.0	0	66.8	0.002
马口鱼	区域5	12.5	7.6	0	79.9	0.003

5.3.4.3　水生生物中的砷形态与营养级的关系

水生生物中的有机砷与营养级（$\delta^{15}N$）的关系如图 5.11 所示。在本研究中，淡水水生生物中无机砷含量范围较广，从未检测出到 0%～87%，显著高于海洋生物。Kucuksezgin 等人（2014）研究发现，Izmir Bay 海洋生物肌肉组织中无机砷含量范围为 0.11%～11.81%。Krishnakumar 等人（2016）研究发现，阿拉伯湾中海洋生物肌肉组织中无机砷含量范围为 0%～1.8%。从图 5.11 中可以明显看到，水生生物中有机砷的比例会随着营养级的增加而增加。也就是说，处在低营养级的水生生物中无机砷的含量要高于处在相对高营养级的水生生物。这种现象的原因有多种，可能是高营养级生物主要以捕食方式获取砷，而低营养级生物主要从环境中获取砷，环境（水体）中的砷形态主要为无机砷（Zheng et al.，2003；Miyashita et al.，2009；Dovick et al.，2016）。因此，低营养级生物的砷形态中无机砷的比例高于高营养级生物。Maeda 等人（1993）通过室内砷添加实验也发现，虾（*Neocaridina denticulata*）→鲤鱼（*Cyprinus carpio*）中无机砷比例从 89% 下降到 78%。

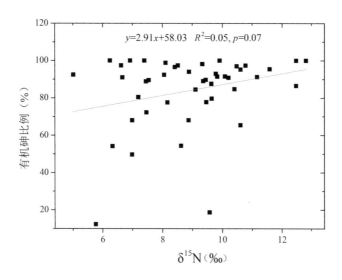

图 5.11 水生生物中的有机砷与营养级（$\delta^{15}N$）的关系

The relationship between organic As and trophic level of aquatic organisms

水生生物中的甲基砷与营养级（$\delta^{15}N$）关系如图 5.12 所示。从图 5.11 中可以明显看到，水生生物中甲基砷的比例会随着营养级的增加而显著增加，推测可能是高级生物具有更高的转化砷的能力。Suhendrayatna 等人（2001a）通过室内砷添加实验发现，当暴露在 As(Ⅲ)水条件下时，虾（*Neocaridina denticulata*）中可检测到 DMA（7%～32%），高营养级肉食性鱼类 *Tilapia mossambica* 中可检测到 TMA（12%～72%），也就是说，高等生物具有更高的甲基化砷的能力。Maeda 等人（1993）通过室内砷添加实验也发现，虾（*Neocaridina denticulata*）中的 DMA 和 TMA 的含量分别为 5.2%和 5.2%，而高营养级肉食性鲤鱼（*Cyprinus carpio*）中的 DMA 和 TMA 含量升高，分别为 7.7%和 13.5%。Kuroiwa 等人（1994）通过室内砷添加实验也发现，虾（*Neocaridina denticulata*）中的 DMA 和 TMA 的含量分别为 36.1%和 3.0%，而高营养级肉鳉鱼（*Oryzias latipes*）中的 TMA 含量明显升高（80%）。

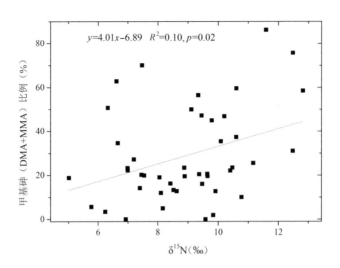

图 5.12 水生生物中的甲基砷与营养级（$\delta^{15}N$）关系

The relationship between methylated As and trophic level of aquatic organisms

5.3.5 石门雄黄矿食用水生生物的健康风险评估

人群通过食用水生生物每日摄入砷的健康风险计算方法如下：

$$EDI = C_{As} \times IR / BW$$

式中，

EDI（μg/kg $_{BW}$/day）为通过食物进入人体的日平均摄入总砷；

C_{As} 为食物砷含量（mg/kg，湿重）；

IR 为每日对某食物的食用量（g/day，湿重）；

BW 为人的体重。

以 THQ 表征由食物引起的砷暴露风险指数：

$$THQ = EDI / RfD$$

式中，

RfD 为砷参考暴露计量（μg/kg BW/day）。USEPA 推荐的无机砷的 RfD 为 0.3 μg/kg BW/day（USEPA，2000）。WHO 推荐的总砷耐受量为 2.14 μg/kg BW/day（WHO，1983）。当 THQ<1 时，表明没有明显的健康风险。但是，当 THQ>1 时，表明可引起人体健康风险，特别是对敏感人群，如孕妇和小孩（Hosseini et al.，2015）。

大量的研究表明，砷的摄入会引发一系列的疾病，如皮肤病、高血压、心血管疾病、糖尿病及生殖系统损伤等，最终导致各种癌症（Sharma and Sohn，2009；Shen et al.，2016）。因此，人体摄入砷的致癌风险研究也非常重要，本研究采用 ILCR 表征砷的致癌风险：

$$\text{ILCR} = \text{EDI} \times q^*$$

式中，

q^* 为无机砷的致癌因子。

USEPA 综合风险信息系统（IRIS）（1998 年最新修订）指出，无机砷引发的皮肤癌的 q^* 值为 1.5 mg/kg BW/day（USEPA，1998）。2010 年，USEPA 根据流行病文献提出了膀胱癌和肺癌的 q^* 值为 25.7 mg/kg BW/day（USEPA，2010）。

本研究采集的水生生物（主要为鱼类）都是当地居民常见的食物。在本研究中，淡水鱼的含水率按 75%计算（Ikemoto et al.，2008；张楠，2013；Cott et al.，2016；Liu et al.，2018），假设每人平均每天食用淡水鱼肌肉组织 100 g，成人体重按 60 kg 计算，石门雄黄矿及周边地区居民砷摄入的风险评估结果如表 5.5 所示。不同区域的总砷含量（湿重）差别很大，区域 4 和区域 5 鱼类肌肉组织中的砷含量最高。居民平均每日食用 100 g 鱼类，每日摄入的总砷为 0.25~8.97 μg kg$_\text{BW}^{-1}$day^{-1}，无机砷含量为 0~4.52 μg kg$_\text{BW}^{-1}$day^{-1}。根据 WHO 推荐的总砷耐受量为 2.14 μg kg$_\text{BW}^{-1}$day^{-1}（WHO，1983）。区域

2、3、5、7 居民每日砷摄入量为 0.25～1.93 μg kg $_{BW}^{-1}$day^{-1}，低于推荐的 2.14 μg kg $_{BW}^{-1}$day^{-1}。但是，区域 4 居民每日砷摄入量为 1.70～8.97 μg kg $_{BW}^{-1}$ day^{-1}，因此存在明显的健康风险。

表 5.5 石门雄黄矿及周边地区居民砷摄入的风险评估结果

The estimated health risk of As by consumption of freshwater fish

		含量（湿重）	EDI	THQ	ILCR[①]	ILCR[②]
区域 2	总砷	0.15～0.83	0.25～0.80	0.12～0.37[③]		
	无机砷	0～0.074	0～0.12	0～0.41	0～6×10^{-4}	0～1×10^{-2}
区域 3	总砷	0.29～2.13	0.48～1.28	0.23～0.60[③]		
	无机砷	0～0.059	0～0.10	0～0.33	0～5×10^{-4}	0～8×10^{-3}
区域 4	总砷	1.02～5.38	1.70～8.97	0.79～4.19[③]		
	无机砷	0.05～2.71	0.08～4.52	0.28～15.06	4×10^{-4}～2×10^{-2}	7×10^{-3}～0.39
区域 5	总砷	0.23～3.67	0.38～1.93	0.18～0.90[③]		
	无机砷	0～0.36	0～0.60	0～2.00	0～3×10^{-3}	0～5×10^{-2}
区域 7	总砷	0.16～0.56	0.27～0.93	0.12～0.44[③]		
	无机砷	0～0.085	0～0.14	0～0.47	0～7×10^{-4}	0～1×10^{-2}

注：①皮肤癌；②膀胱癌和肺癌；③EDI 与每日砷耐受量的比值。

在本研究中，鱼类无机砷可占提取砷的 0%～50.3%，而无机砷的毒性远高于有机砷，因此有必要对鱼类中的无机砷摄入的健康风险进行评估。不同区域无机砷含量（湿重）差别也很大，区域 2、区域 3 和区域 7 鱼类肌肉组织中的无机砷含量<0.085 mg/kg，而区域 4 和区域 5 鱼类肌肉组织中的无机砷含量最高可达 2.71 mg/kg。USEPA 推荐的无机砷的 RfD 为 0.3 μg/kg $_{BW}^{-1}$day^{-1}（USEPA，2000）。计算可得，区域 2、区域 3 和区域 7 的 THQ 均<0.5，因此不存在明显的健康风险。但是区域 4 和区域 5 的 THQ 整体较高，说明食用区域 4 和区域 5 的鱼类可能会引起人体明显的健康风险。本研究中砷暴露的致癌风险 ILCR 基本高于推荐值 1×10^{-5} 和最高耐受值 1×10^{-4}，说明食用本地区的鱼类存在一定的致癌风险。

5.4　本章小结

本章主要分析石门雄黄矿水生生态系统（主要为水生动物）对砷的富集及其形态特征。石门雄黄矿过去的砷采矿活动对环境中的砷污染产生了显著的影响，成为水生食物网中砷的一个来源。结果表明，水生生物中的砷含量范围为 0.60～45.75 mg/kg，水体砷污染程度、组织和种类差异性、体重、体长、栖息水层和食性等都会影响水生生物对砷的富集。不同区域的水生生物肌肉组织中的 $\delta^{15}N$ 范围为 4.31‰～12.98‰，表现为滤食性<植食性<杂食性<肉食性，与生物食物链营养级一致。另外，随着水生生物的营养级（$\delta^{15}N$）的增加，水生生物对砷的富集量和富集系数呈现下降趋势，但 $\delta^{15}N$ 与总砷并无显著的相关性，表明在食物网中砷并无生物放大效应。采用 HPLC-ICP-MS 进行砷形态分析结果表明，水生生物中以有机砷为主，并出现大量的未知砷。研究发现，随着水生生物体内砷含量的增加，其无机砷 As(III) 和 As(V) 的浓度和比例显著增加。这种相关性可能是水生生物对无机砷有高的同化和富集能力，或者对无机砷有低的转化能力所致。我们对采集的淡水鱼类进行健康风险评估，发现区域 4 和区域 5 的鱼类具有较高的健康风险，需要引起重视。

参 考 文 献

耿頔, 韦朝阳, 季宏兵. 湖泊水体悬浮物中痕量砷的测定方法[J]. 环境科学学报, 2015, 35:1728-1734.

王玉玉, 于秀波, 张亮, 等. 应用碳、氮稳定同位素研究鄱阳湖枯水末期水生食物网结构[J]. 生态学报, 2009, 29:1181-1188.

张楠, 韦朝阳, 杨林生. 淡水湖泊系统中砷的赋存与转化行为研究进展[J]. 生态学报, 2013, 33:337-347.

ALAMDAR A, EQANI SAMAS, HANIF N, et al. Human exposure to trace metals and arsenic via consumption of fish from river Chenab, Pakistan and associated health risks[J]. Chemosphere, 2017, 168:1004-1012.

ARAIN MB, KAZI TG, JAMALI MK, et al. Total dissolved and bioavailable elements in water and sediment samples and their accumulation in *Oreochromis mossambicus* of polluted Manchar Lake[J]. Chemosphere, 2008, 70:1845-1856.

AZIZUR RAHMAN M, HASEGAWA H, LIM RP. Bioaccumulation, biotransformation and trophic transfer of arsenic in the aquatic food chain[J]. Environ. Res., 2012, 116:118-135.

CAUMETTE G, KOCH I, ESTRADA E, et al. Arsenic speciation in plankton organisms from contaminated lakes: transformations at the base of the freshwater food chain[J]. Environ. Sci. Technol., 2011, 45:9917-9923.

CAUMETTE G, KOCH I, MORIARTY M, et al. Arsenic distribution and speciation in *Daphnia pulex*[J]. Sci. Total Environ., 2012, 432:243-250.

CHEN CY, FOLT CL. Bioaccumulation and diminution of arsenic and lead in a freshwater food web[J]. Environ. Sci. Technol., 2000, 34:3878-3884.

CIARDULLO S, AURELI F, RAGGI A, et al. Arsenic speciation in freshwater fish: Focus on extraction and mass balance[J]. Talanta, 2010, 81:213-221.

COTT PA, ZAJDLIK BA, PALMER MJ, et al. Arsenic and mercury in lake whitefish and burbot near the abandoned Giant Mine on Great Slave Lake[J]. J Great Lakes Res., 2016, 42:223-232.

CUI B, ZHANG Q, ZHANG K, et al. Analyzing trophic transfer of heavy metals for food webs in the newly-formed wetlands of the Yellow River Delta, China[J]. Environ Pollut., 2011, 159:1297-1306.

CULIOLI JL, FOUQUOIRE A, CALENDINI S, et al. Trophic transfer of arsenic and antimony in a freshwater ecosystem: A field study[J]. Aquat Toxicol, 2009, 94:286-293.

DE ROSEMOND S, XIE Q, LIBER K. Arsenic concentration and speciation in five freshwater fish species from Back Bay near Yellowknife, NT, Canada[J]. Environ Monit Assess, 2008, 147:199-210.

DHANEESH KV, NOUSHAD KM, KUMAR. Nutritional evaluation of commercially important fish species of Lakshadweep Archipelago, India[J]. PLoS ONE, 2012, 7:e45439.

DOVICK MA, KULP TR, ARKLE RS, et al. Bioaccumulation trends of arsenic and antimony in a freshwater ecosystem affected by mine drainage[J]. Environ Chem, 2016, 13:149-159.

FARAG AM, NIMICK DA, KIMBALL BA, et al. Concentrations of metals in water, sediment, biofilm, benthic macroinvertebrates, and fish in Boulder River watershed, Montana, and the role of colloids in metal uptake[J]. Arch Environ Contam Toxicol, 2007, 52:397-409.

FLIEDNER A, RÜDEL H, KNOPF B, et al. Spatial and temporal trends of metals and arsenic in German freshwater compartments[J]. Environ Sci Pollut Res, 2014, 21:5521-5536.

FRY B. Food web structure on Georges Bank from stable C, N, and S isotopic compositions[J]. Limnol. Oceanogr, 1988, 33:1182-1190.

FU Z, WU F, MO C, et al. Bioaccumulation of antimony, arsenic, and mercury in the vicinities of a large antimony mine, China[J]. Microchem J, 2011, 97:12-19.

FU Z, WU F, AMARASIRIWARDENA D, MO C, LIU B, ZHU J, DENG Q, LIAO H. Antimony, arsenicandmercury in the aquatic environmentandfish in a large antimony mining area in Hunan, China[J]. Sci. Total Environ, 2010, 408, 3403-3410.

GEDIK K, KONGCHUM M, DELAUNE RD, et al. Distribution of arsenic and other metals in crayfish tissues (*Procambarus clarkii*) under different production practices[J]. Sci Total Environ, 2017, 574:322-331.

GRINHAM A, KVENNEFORS C, FISHER PL, et al. Baseline arsenic levels in marine and terrestrial resources from a pristine environment: Isabel Island, Solomon Islands[J]. Mar Pollut Bull, 2014, 88:354-360.

HAS-SCHÖN E, BOGUT I, VUKOVIĆ B, et al. Distribution and age-related bioaccumulation of lead (Pb), mercury (Hg), cadmium (Cd), and arsenic (As) in tissues of common carp (*Cyprinus carpio*) and European catfish (*Sylurus glanis*) from the Buško Blato reservoir (Bosnia and Herzegovina)[J]. Chemosphere, 2015, 135, 289-296.

HELSEN L. Sampling technologies and air pollution control devices for gaseous and particulate arsenic: A review[J]. Environ Pollut, 2005, 137:305-315.

HIRATA S, TOSHIMITSU H. Determination of arsenic species and arsenosugars in marine samples by HPLC-ICP-MS[J]. Appl Organometal Chem, 2007, 21:447-454.

HONG S, KHIM JS, PARK J, et al. Species- and tissue-specific bioaccumulation of arsenicals in various aquatic organisms from a highly industrialized area in the Pohang City, Korea[J]. Environ Pollut, 2014, 192:27-35.

HOSSEINI SM, KARAMINASAB M, BATEBINAVAEI M, et al. Assessment of the essential elements and heavy metals content of the muscle of Kutum (*Rutilus frisii kutum*) from the south Caspian Sea and potential risk assessment[J]. Iran J Fish Sci, 2015, 14:660-671.

IKEMOTO T, TU NPC, OKUDA N, et al. Biomagnification of trace elements in the aquatic food web in the Mekong Delta, South Vietnam using stable carbon and nitrogen isotope analysis[J]. Arch Environ Contam Toxicol, 2008, 54:504-515.

JANKONG P, CHALHOUB C, KIENZL N, et al. Arsenic accumulation and speciation in freshwater fish living in arsenic-contaminated waters[J]. Environ Chem, 2007, 4:11-17.

JUNCOS R, ARCAGNI M, RIZZO A, et al. Natural origin arsenic in aquatic organisms from a deep oligotrophic lake under the influence of volcanic eruptions[J]. Chemosphere, 2016, 144:2277-2289.

KIM JH, KANG JC. The arsenic accumulation and its effect on oxidative stress responses in juvenile rockfish, *Sebastes schlegelii*, exposed to waterborne arsenic (As^{3+}) [J]. Environ Toxicol Phar, 2015, 39:688-676.

KRISHNAKUMAR PK, QURBAN MA, STIBOLLER M, et al. Arsenic and arsenic species in shellfish and finfish from the western Arabian Gulf and consumer health risk assessment[J]. Sci Total Environ, 2016.

KUCUKSEZGIN F, GONUL LT, TASEL D. Total and inorganic arsenic levels in some marine organisms from Izmir Bay (Eastern Aegean Sea): A risk assessment[J]. Chemosphere, 2014, 112:311-316.

KUROIWA T, OHKI A, NAKA K, et al. Biomethylation and biotransformation of arsenic in a freshwater food chain: Green alga (*Chlorella vulgaris*)→shrimp (*Neocaridina denticulata*)→killifish (*Oryzias iatipes*)[J]. Appl Organometal Chem, 1994, 8:325-333.

LIU Y, LIU G, YUAN Z, et al. Heavy metals (As, Hg and V) and stable isotope ratios ($\delta^{13}C$ and $\delta^{15}N$) in fish from Yellow River Estuary, China[J]. Sci Total Environ, 2018.

MAEDA S, OHKI A, KUSADOME K, et al. Bioaccumulation of arsenic and its fate in a freshwater food chain[J]. Appl Organometal Chem, 1992, 23:213-219.

MAEDA S, MAWATARI K, OHKI A, et al. Arsenic metabolism in a freshwater food chain: Blue–green alga (*Nostoc* sp.)→shrimp (*Neocaridina denticulata*)→carp (*Cyprinus carpio*)[J]. Appl Organometal Chem, 1993, 7:467-476.

MAHER W, GOESSLER W, KIRBY J, et al. Arsenic concentrations and speciation in the tissues and blood of sea mullet (*Mugil cephalus*) from Lake Macquarie NEW, Australia[J]. Mar Chem, 1999, 68:169-182.

MAHER W, FOSTER S, KRIKOWA F. Arsenic species in Australian temperate marine food chains[J]. Mar Freshwater Res, 2009, 60:885-892.

MANDAL BK, SUZUKI KT. Arsenic round the world: A review[J]. Talanta, 2002, 58:201-235.

MARCHESE MR, SAIGO M, ZILLI FL, et al. Food webs of the Paraná

River floodplain: Assessing basal sources using stable carbon and nitrogen isotopes[J]. Limnologica, 2014, 46:22-30.

MIYASHITA S, SHIMOYA M, KAMIDATE Y, et al. Rapid determination of arsenic species in freshwater organisms from the arsenic-rich Hayakawa River in Japan using HPLC-ICP-MS[J]. Chemosphere, 2009, 75:1065-1073.

MOGREN CL, WALTON WE, PARKER DR, et al. Trophic transfer of arsenic from an aquatic insect to terrestrial insect predators[J]. Plos One, 2013, 8:1-6.

MORIARTY MM, KOCH I, GORDON RA, et al. Arsenic speciation of terrestrial invertebrates[J]. Environ Sci Technol, 2009, 43:4818-4823.

NG JC. Environmental contamination of arsenic and its toxicological impact on humans[J]. Environ Chem, 2005, 2:146-160.

PEREIRA AA, VAN HATTUM B, DE BOER J, et al. Trace elements and carbon and nitrogen stable isotopes in organisms from a tropical coastal lagoon[J]. Arch Environ Contam Toxicol, 2010, 59:464-477.

PERERA PACT, SUNDARABARATHY TV, SIVANANTHAWERL T, et al. Arsenic and cadmium contamination in water, sediments and fish is a consequence of paddy cultivation: Evidence of river pollution in Sri Lanka[J]. Achievements in the Life Sciences, 2016, 10:144-160.

PHAN K, STHIANNOPKAO S, Heng S, et al. Arsenic contamination in the food chain and its risk assessment of populations residing in the Mekong River basin of Cambodia[J]. J Hazard Mater, 2013, 262:1064-1071.

POST DM. Using stable isotopes to estimate trophic position: Models, methods, and assumptions[J]. Ecology, 2002b, 83:703-718.

REVENGA JE, CAMPBELL LM, ARRIBÉRE MA, et al. Arsenic, cobalt and chromium food web biodilution in a Patagonia mountain lake[J]. Ecotox

Environ Safe, 2012, 81:1-10.

ROIG N, SIERRA J, ORTIZ JD, et al. Integrated study of metal behavior in Mediterranean stream ecosystems: A case-study[J]. J Hazard Mater, 2013, 263P:122-130.

RUTTENS A, BLANPAIN AC, TEMMERMAN LD, et al. Arsenic speciation in food in Belgium, Part 1: Fish, molluscs and crustaceans[J]. J Geochem Explor, 2012, 121:55-61.

SAIGO M, ZILLI FL, MARCHESE MR, et al. Trophic level, food chain length and omnivory in the Paraná River: A food web model approach in a floodplain river system[J]. Ecol Res, 2015, 30:843-852.

SCHAEFFER R, FRANCESCONI KA, KIENZL N, et al. Arsenic speciation in freshwater organisms from the river Danube in Hungary[J]. Talanta, 2006, 69:856-865.

SHAH AQ, KAZI TG, ARAIN MB, et al. Hazardous impact of arsenic on tissues of same fish species collected from two ecosystem[J]. J Hazard Mate, 2009a, 167:511-515.

SHAH AQ, KAZI TG, ARAIN MB, et al. Comparison of electrothermal and hydride generation atomic absorption spectrometry for the determination of total arsenic in broiler chicken[J]. Food Chem, 2009b, 113:1351-1355.

SHARMA VK, SOHN M. Aquatic arsenic: Toxicity, speciation, transformations, and remediation[J]. Environ Int, 2009, 4:743-759.

SHEN H, NIU Q, XU, MC, et al. Factors affecting arsenic methylation in arsenic-exposed humans: A systematic review and meta-analysis[J]. Int J Env Res Pub He, 2016, 13:205-222.

SHIOMI K, SUGIYAMA Y, SHIMAKURA K, et al. Arsenobetaine as the major arsenic compound in the muscle of two species of freshwater fish[J]. Appl Organomet Chem, 1995, 9:105-109.

ŠLEJKOVEC Z, BAJC Z, DOGANOC DZ. Arsenic speciation patterns in freshwater fish[J]. Talanta, 2004, 62:931-936.

SMEDLEY PL, KINNIBURGH DG. A review of the source, behaviour and distribution of arsenic in natural waters[J]. Appl Geochem, 2002, 17, 517-568.

SUHENDRAYATNA, OHKI A, MAEDA S. Biotransformation of arsenite in freshwater food chain models[J]. Appl Organometal Chem, 2001a,15:277-284.

SUHENDRAYATNA, OHKI A, NAKAJIMA T, et al. Metabolism and organ distribution of arsenic in the freshwater fish *Tilapia mossambica*[J]. Appl Organometal Chem, 2001b,15:566-571.

SUHENDRAYATNA, OHKI A, NAKAJIMA T, et al. Studies on the accumulation and transformation of arsenic in freshwater organisms Ⅱ. Accumulation and transformation of arsenic compounds by *Tilapia mossambica*[J]. Chemosphere, 2002, 46:325-331.

TANG JW, LIAO YP, YANG ZH, et al. Characterization of arsenic serious-contaminated soils from Shimen realgar mine area, the Asian largest realgar deposit in China[J]. J Soil Sediment, 2016, 16,1519-1528.

USEPA. Arsenic, inorganic[p]. CASRN 7440-38-2, 1998.

USEPA, IRIS. Toxicological Review of Inorganic Arsenic (Cancer)[p]. EPA/635/R-10/001; Washington, DC, 2010.

VANDER ZANDEN MJ, CABANA G, RASMUSSEN JB. Comparing trophic position of freshwater fish caculated using stable nitrogen isotope ratios (δ^{15}N) and literature dietary data[J]. Can J Fish Aquat Sci, 1997,54:1142-1158.

VANDER ZANDEN MJ, CASSELMAN JM, RASMUSSEN JB. Stable

isotope evidence for the food web consequences of species invasion in lakes[J]. Nature, 1999, 401:464-467.

VIZZINI S, COSTA V, TRAMATI C, et al. Trophic transfer of trace elements in an isotopically constructed food chain from a semi-enclosed marine coastal area (Stagnone di Marsala, Sicily, Mediterranean)[J]. Arch Environ Contam Tox, 2013, 65:642-653.

WANG S, LI B, ZHANG M, et al. Bioaccumulation and trophic transfer of mercury in a food web from a large, shallow, hypereutrophic lake (Lake Taihu) in China[J]. Environ Sci Pollut Res, 2012, 19:2820-2831.

WATANABE K, MONAGHAN MT, TAKEMON Y, et al. Biodilution of heavy metals in a stream macroinvertebrate food web: Evidence from stable isotope analysis[J]. Sci Total Environ, 2008, 394:57-67.

WHO. Evaluation of certain food additives and contaminants (Twenty-seventh report of the joint FAO/WHO expert committee on food additives)[R]. World Health Organization technical report Series No. 696, 1983.

WILLIAMS G, WEST JM, KOCH I, et al. Arsenic speciation in the freshwater crayfish, *Cherax destructor* Clark [J]. Sci Total Environ, 2009, 407:2650-2658.

YANG F, GENG D, WEI C, et al. Distribution of arsenic between the particulate and aqueous phases in surface water from three freshwater lakes in China[J]. Environ Sci Pollut Res, 2016, 23:7452-7461.

YANG F, ZHANG N, WEI C, et al. Arsenic speciation in organisms from two large shallow freshwater lakes in China[J]. Bull Environ Contam Toxicol, 2017, 98:226-233.

ZHANG W, WANG WX. Large-scale spatial and interspecies differences in trace elements and stable isotopes in marine wild fish from Chinese waters[J]. J Hazard Mater, 2012, 215-216:65-74.

ZHANG N, WEI C, YANG L. Occurrence of arsenic in two large shallow freshwater lakes in China and a comparison to other lakes around the world[J]. Microchem J, 2013,110:169-177.

ZHENG J, HINTELMANN, DIMOCK B, et al. Speciation of arsenic in water, sediment, and plants of the Moira watershed, Canada, using HPLC coupled to high resolution ICP-MS[J]. Anal Bioanal Chem, 2003, 377:14-24.

ZHU YM, WEI CY, YANG LS. Rehabilitation of a tailing dam at Shimen County, Hunan Province: Effectiveness assessment[J]. Acta Ecologica Sinica, 2010, 30:178-183.

ZHU X, WANG R, LU X, et al. Secondary minerals of weathered orpiment-realgar-bearing tailings in Shimen carbonate-type realgar mine, Changde, Central China[J]. Miner Petrol, 2015,109:1-15.

第 6 章

食物来源影响水生生物富集和转化重金属的差异性

第6章 食物来源影响水生生物富集和转化重金属的差异性

6.1 食物来源影响水生生物富集和转化重金属的研究意义

环境重金属污染具有隐蔽性、长期性和不可逆性的特点，因此被广泛关注（Bailey et al.，1999；Li et al.，2014）。重金属污染的主要来源包括自然来源和人为来源，其中矿产的开采及其相应的工业活动、废水处理和排放是重金属的主要来源（Wei and Yang，2010；Xiao et al.，2022）。排放到水生环境中的重金属会损害水生生物，导致随后在高级营养捕食者中积累，并通过食物链对人类构成潜在的健康风险。

砷是地表环境中普遍存在的微量元素，对动植物有广泛的致毒性（Mandal and Suzuki，2002）。砷对生物体的毒性与砷的浓度、形态、生物有效性及摄取方式有关（Hughes，2002；Mandal and Suzuki，2002；Azizur Rahman et al.，2012）。因此，对环境和生物体中砷含量及存在形态进行分析具有重要意义。水生生物可以通过多种渠道摄入食物和水、沉积物中的砷，并在体内富集，最终进入人体，引发一系列的健康问题（Dhaneesh et al.，2012；Alamdar et al.，2017）。近年来，由于水体砷污染的频繁报道，水生生态系统中的砷研究已引起人们的高度关注（张楠 等，2013）。到目前为止，我们对于海水环境中砷的分布及生物体中的砷含量与形态特征已有比较全面的认识，但是关于淡水水生生物中砷特征的报道并不多见（Azizur Rahman et al.，2012）。

淡水鱼类中砷含量相对较低，一般为 0.1～10 mg/kg，远低于海洋生物中的砷含量（Mandal and Suzuki，2002）。研究表明，淡水鱼类中的砷形态非常复杂，并非单一砷形态占绝对优势，通常由多种砷形态组成，不同物种间的砷形态差异较大（de Rosemond et al.，2008；Hong et al.，2014；Jia et al.，2018）。但是，我们对于水生生物对砷的富集和转化的种间差异性的机制研究还知之甚少。因此，更好地了解重金属的营养转移，对于维持健康的生态系统和保护公众健康具有重要意义（O'Mara et al.，2019）。

水生生物吸收重金属的两个主要途径是：通过食物（水体中的悬浮颗粒物、浮游生物及被捕食者等）吸收累积重金属（Wang and Fisher，1999；Mathews and Fisher，2009）；通过鱼鳃等组织中亲脂性细胞膜长时间累积水体中的溶解态重金属（Azizur Rahman et al.，2012；Wang and Fisher，1999；Pouil et al.，2018）。然而，在过去的几十年中，大多数的重金属研究集中在水生生物的水相吸收上。有学者认为，水相中的主要砷形态为无机砷 As(V)，水生生物从水相中累积的 As(V) 很少，而直接吸附在生物体表面组织上的砷又很容易释放到周围的水体中，因此水生生物砷的水相暴露并不重要（Fowler and Ünlü，1978；Ventura-Lima et al.，2011；Culioli et al.，2009）。一些室内模拟实验也证实，食物相暴露是水生生物砷及其他重金属累积的主要途径（Pedlar and Klaverkamp，2002；Williams et al.，2010；Erickson et al.，2011，2019；Mathews and Fisher，2009；Erickson et al.，2011；O'Mara et al.，2019；McDonald et al.，2021）。研究表明，在 *Sillago ciliate* 和 *Ambassis jacksoniensis* 体内基本检测不到溶解性的 Cd 和 Zn，而 *A. jacksoniensis* 和 *Acanthopagrus schlegeli* 体内大于 95% 的 Cd 和 Zn 主要

第6章 食物来源影响水生生物富集和转化重金属的差异性

通过食物相富集(Zhang and Wang,2007;Creighton and Twining,2010;O'Mara et al.,2019)。一些淡水鱼类中的砷形态主要为 AsB,这主要与其食物相暴露有关(Šlejkovec et al.,2004;Soeroes et al.,2005;Schaeffer et al.,2006;Ciardullo et al.,2010)。室内实验主要在短时间内对一种或两种食物(如颗粒相、虾和藻类)进行研究,而野生水生生物的食物来源多种多样(Creighton and Twining,2010)。因此,量化水生生物的潜在食物来源对于理解淡水生态系统中重金属的营养转移和控制重金属污染至关重要(McDonald et al.,2021)。

食物组成和各食物组分中的砷赋存特征等多种因素都会影响水生生物对砷的富集和转化。已有研究表明,与汞不同,而与其他重金属元素相似,砷在其营养转化过程中呈现生物浓缩的趋势,而摄食习性差异可能是鱼类累积砷差异的主要原因(Chen and Folt,2000;Cui et al.,2011;Vizzini et al.,2013;Revenga et al.,2012;Foust et al.,2016)。Burger 等人(2002)对美国萨瓦纳河 11 种淡水鱼类中的砷研究发现,从低营养级到高营养级鱼类中的总砷含量由 0.32 mg/kg 下降到 0.03 mg/kg。Chen 和 Folt(2000)对美国 Upper Mystic 湖的研究发现,以浮游生物为食的鱼类中的砷含量大于高营养级的肉食性鱼类。de Rosemond 等人(2008)发现,加拿大大奴湖贝克湾中的水生生物中,杂食性鱼类中的砷含量高于肉食性鱼类。但是,这些研究都是基于水生生物的传统食性划分的,缺乏对其食物组成的定性定量分析。国内外关于鱼类摄食的研究已经非常普遍,一般来说鱼类的捕食范围广泛,通常由多种生物饵料组成。因此,目前的研究不能全面、准确反映摄食习性对水生生物富集砷的影响机制。

尽管已有一些对水生生物砷富集传递作用的报道,但尚未有研究关注不同食源水生生物体中砷的生物转化行为。浮游生物是水生生物重要的食物来源,研究表明,在淡水湖泊中,浮游植物中的砷以无机砷形态存在,大于92%的砷为 As(V),而浮游动物中除无机砷外,还含有少量的甲基砷、砷糖及 AsB（Caumette et al.，2011，2012）。鱼类摄入食物中的砷并在体内转化 As(Ⅴ)为 As(Ⅲ)或甲基化形态，可能涉及多种机制。砷的甲基化是砷同生物细胞中蛋白质结合的过程，改变酶蛋白和非酶蛋白的结构和活性，进而对生物体产生毒害作用（Shen et al.，2013；Ghosh et al.，2013）。甲基砷可在生物体内通过对无机砷的生物转化作用产生，并缓慢排到体外（Azizur Rahman et al.，2012）。一般区域的水生生物及其食物相砷含量相对较低，难以满足检测限的要求。目前，关于淡水鱼类对食物中砷的转化研究的文献报道比较零星，多基于室内模拟实验研究。Maeda 等人（1993）通过室内砷添加实验发现，藻类（*Nostoc sp.*）→虾（*Neocaridina denticulata*）→肉食性鲤鱼（*Cyprinus carpio*）中的甲基砷含量从<1%增加至 21%。Kuroiwa 等人（1994）通过室内砷添加实验发现，藻类（*Chlorella vulgaris*）→虾（*Neocaridina denticulata*）→鳉鱼（*Oryzias latipes*）中的甲基砷含量从 10.9%连续增加至 80%。Suhendrayatna 等人（2001）通过室内砷添加实验发现，藻类（*Chlorella vulgaris*）→浮游动物（*Daphnia magna*）→虾（*Neocaridina denticulata*）→肉食性鱼类 *Tilapia mossambica* 中甲基砷含量也连续增加。但是，由于室内模拟可研究的生物种类单一，食物相暴露简单，且暴露时间短，还受到实验条件等诸多限制，无法反映野外复杂生存环境下水生生物砷富集转化与食物相之间的真实情况。因此，自然条件下食物来源对水生生物体转化砷的差异性及其机制还需要开展更多的研究加以回答。

值得注意的是，水生生物在生长发育过程中，其摄食规律（食源和食性）会发生转变，如当某些水生生物的栖息地、季节等发生变化时，其摄食对象也会发生改变，由此影响砷的生物效应（Cui et al.，2011；余杨 等，2013；韦丽丽 等，2016）。草鱼作为我国主要的淡水养殖经济鱼类之一，其食性在早期生长发育阶段主要以浮游动物为食，在体长达到 50 mm 时转变为食草性（张金梅，1982）。徐军（2005）研究发现，巢湖小湖鲚的食物以浮游性为主，而大湖鲚的摄食方式更偏向于底栖食性和肉食性。

Genner 等人（2003）发现，Malawi 湖中 *Pseudotropheus callainos* 在生长发育过程中，食性从浮游性到底栖性转变。不同的季节，水温、光照等会发生很大的改变，水生生物的饵料生物受环境影响，最终导致淡水鱼类的食性随季节的变化而改变（刘其根 等，2015）。另外，不同生物对砷的耐受性不同，当生境中砷污染程度升高时，对砷毒性敏感的物种数量会减少甚至消失，而耐砷毒性的物种数量则会增加。Moore 等人（1979）研究发现，当水体砷浓度高达 2000 μg/L 时，沉积物中的寡毛类物种将会消失。Mori 等人（1999）对法国科西嘉岛 Corsican 河的研究发现，水体及沉积物中的砷污染会导致雄黄矿下游的水蛭、浮游类及寡毛类底栖物种消失；相反，水生石蝇稚虫及腹足类软体动物却有所增加。水环境中的有机质分为内源性和外源性两种来源，但是当环境受到外界干扰时，不同类型的有机质丰度也会发生变化，从而直接影响水生生物的食物来源（Watanabe et al.，2008）。总而言之，鱼类的体长、季节、栖息水域等都会直接影响水生生物获取食物来源的可得性或直接导致水生生物食性转化。因此，研究水生生物中各食物组成及其变化规律对生物砷行为研究具有非常重要的意义。

食物组成是水生生物重金属富集和转化的重要影响因素。传统研究鱼类食物组成的方法主要是胃内容物分析，该方法具有更直观、即时性等优点。同时，胃内容物分析结果可直接用于研究鱼类摄食随发育阶段和时空的变化规律。但是，其本身存在着许多不足。胃内容物分析只能确定被捕前短期内的食性，且胃内容物中通常存在的是难以消化的食物，同时在研究小型动物方面存在困难，影响研究人员对食物的鉴定（Beaudoin et al.，1999；王玉玉 等，2009）。同时，胃内容物分析需要大量的样本在广泛的地理和时间尺度上跟踪饮食（Varela et al.，2019）。近年来，稳定同位素比值能够反映研究对象的长期消化吸收的食物来源，弥补了胃内容物分析的不足，是近年来用于研究动物食性的有效手段（Overman and Parrish，2001；Philips et al.，2005；Vander Zanden and Rasmussen，2001；Ai et al.，2019）。

为了定量估计每个潜在食物来源的贡献，科研人员已经相继开发了多个混合模型，包括 IsoSource、SIAR 和 MixSIR（Philips and Gregg，2003；Moore and Semmens，2008；Parnell et al.，2010）。然而，许多研究人员并不确定这些混合模型的正确应用和解释。为了解决这个问题，我们利用贝叶斯稳定同位素混合模型（MixSIAR）来确定生态环境中食物来源的比例贡献（Stock et al.，2018）。虽然稳定同位素技术克服了传统胃内容物分析的一些缺陷，但只能确定生物主要的食物组成及贡献比例范围，不能全面准确地判断环境变化下水生生物的食物组成。因此，往往需要胃内容物分析作为辅助，两种方法结合使用，可以研究水生生物完整的摄食习性。

石门雄黄矿土壤中的砷含量可达到 5240 mg/kg，水中砷含量可达到 40.1 mg/L（Zhu et al.，2015；Tang et al.，2016）。附近的黄水溪是当地居民的重要水源（Zhu et al.，2010；Li et al.，2020）。然而，许多研究只关注砷，而对当地居民与水生生物和食物来源中的其他重金属的关系的研究很少。本研究的目的是：①测定水生生物及其潜在食物来源中 8 种重金属含量，包括 As、Cd、Pb、Cr、Cu、Mn、Ni 和 Zn；②通过碳氮同位素、MixSIAR 模型结合胃内容物分析，估算不同潜在食物来源对重金属的比例贡献；③评价重金属从潜在食物来源到水生生物的生物放大或生物转化。我们的研究结果为进一步了解淡水生态系统中重金属的传递过程，为当地渔业的管理及维护健康的生态系统和生物多样性提供了重要的信息。

6.2 砷矿区水生环境生物样品采集、总砷及砷形态分析方法

近几十年来，废弃石门雄黄矿有大量尾矿、砷渣和砷灰排放到黄水溪上游（Tang et al., 2016），因此，本研究选取石门雄黄矿黄水溪作为研究区域。2020 年 9 月，根据我们之前的现场调查观察到的污染水平（见第 3 章），我们选择黄水溪上游至下游的 4 个点位，包括点位 A、点位 B、点位 C 和点位 D，采集水生生物。简单来说，点位 A 位于矿区，点位 B 位于雄黄矿附近，点位 C 位于尾矿坝附近，点位 D 则位于黄水溪及其支流的汇合处。

在当地渔民的帮助下，我们共收集到马口鱼、泥鳅、沙塘鳢、鲶鱼、鳑鲏、麦穗鱼和青虾 7 种（赵文，2005）。为了检验体型的影响，马口鱼个体被分为三组：马口鱼（S）体长小于 2 cm，马口鱼（M）体长为 2~10cm，马口鱼（L）体长大于 10 cm。沙塘鳢被分为两组：沙塘鳢（M）体长小于 10 cm，沙塘鳢（L）体长大于 10 cm。石门雄黄矿区水生生物样品的相关参数如表 6.1 所示。

我们在黄水溪采集了水生生物的潜在食物来源，其中，体长小于2cm的马口鱼（S）被认为是一种潜在的食物来源。附生岩石相是通过使用牙刷擦洗10~20块岩石的上层表面获得的，所得材料用蒸馏去离子水洗涤后装入聚乙烯袋中（Watanabe et al.，2008）。枯枝落叶的获取方法为，将地表水通过一个0.1mm的尼龙筛后，留在筛子上的样品保存在自封袋中。通过分级过滤收集粗颗粒有机质（粗颗粒相：64μm~0.1mm）和细颗粒有机质（细颗粒相：0.45~64μm）。粗颗粒有机质：用25#浮游生物网过滤后收集颗粒物，静置，除去底部石头等碎屑和上清液，装入自封袋保存。细颗粒有机质：将经25#浮游生物网过滤的水样用0.45μm网过滤，得到膜上的颗粒物质，刮下膜上的物质，实验室内冻干保存。所有组织样品在冷冻干燥和研磨后进行化学分析，化学分析见第2章。

表6.1 石门雄黄矿区水生生物样品的相关参数

Sample details of organisms from Shimen Realgar Mine area

采样点位	生物名称	样品量	体重(g)	体长(cm)	栖息水层	食性	$\delta^{13}C$（‰）	$\delta^{15}N$（‰）
点位A	青虾	3	2~3	5~6	底栖	杂食	-18.65±1.63	10.76±1.05
点位A	泥鳅	2	5~5.5	10~11	底栖	杂食	-16.35±0.86	10.96±0.28
点位A	沙塘鳢（M）	3	4~5	8~9	底栖	肉食	-17.15±0.50	10.39±0.22
点位A	马口鱼（L）	>20	10~15	10~13	上层	肉食	-19.50±0.66	10.27±1.09
点位A	马口鱼（M）	>30	3~9	5~9	上层	杂食	-18.54±0.81	11.06±0.65
点位A	马口鱼（S）	>100	0~2	1~3	上层	杂食	-16.13±0.73	12.38±0.75
点位B	青虾	6	2~4	5~6	底栖	杂食	-18.43±1.05	10.89±0.46
点位B	沙塘鳢（M）	11	1~3	5~7	底栖	肉食	-20.68±0.69	10.15±0.42
点位B	马口鱼（L）	>20	10~19	10~12	上层	肉食	-19.66±0.88	9.40±1.21
点位B	马口鱼（M）	>30	3~9	5~9	上层	杂食	-18.50±0.90	10.96±0.65
点位B	马口鱼（S）	>100	0~2	1~3	上层	杂食	-17.17±1.38	11.78±0.41
点位C	泥鳅	3	7~8.5	10~12	底栖	杂食	-18.62±0.40	10.52±0.80
点位C	马口鱼（L）	>20	10~17	10~13	上层	肉食	-16.05±1.29	11.61±0.62
点位C	马口鱼（M）	>50	3~9	5~9	上层	杂食	-15.54±1.01	11.91±0.35

续表

采样点位	生物名称	样品量	体重(g)	体长(cm)	栖息水层	食性	$\delta^{13}C$ (‰)	$\delta^{15}N$ (‰)
点位 C	马口鱼（S）	>100	0~2	1~3	上层	杂食	−16.29±2.09	11.22±0.51
点位 D	青虾	4	1~2	2.5~3	底栖	杂食	−23.71±2.11	9.03±0.08
点位 D	泥鳅	3	8~17	10~16	底栖	杂食	−22.75±1.24	12.54±0.78
点位 D	鲶鱼	2	20~21	12~13	中层	肉食	−22.81±0.63	8.70±1.05
点位 D	鳑鲏	19	1~4	4~6	上层	杂食	−17.82±1.49	10.68±0.47
点位 D	沙塘鳢（L）	2	10~20	10~12	底栖	肉食	−17.07±1.97	10.99±0.62
点位 D	沙塘鳢（M）	5	1~9	4~8	底栖	肉食	−18.65±0.93	11.93±0.33
点位 D	麦穗鱼	1	3	8	上层	杂食	−17.16±2.41	10.65±0.37
点位 D	马口鱼（L）	>20	10~23	10~13	上层	肉食	−19.78±1.77	11.27±1.35
点位 D	马口鱼（M）	>50	3~9	5~9	上层	杂食	−17.35±2.53	11.34±1.15
点位 D	马口鱼（S）	>100	0~2	1~3	上层	杂食	−17.91±1.68	12.37±1.15

接下来，我们从黄水溪采集的 43 条鱼中收集胃内容物（一些鱼类数量不足无法鉴定），包括马口鱼（点位 A，n=8；点位 B，n=8；点位 C，n=11；点位 D，n=11），泥鳅（点位 C，n=1；点位 D，n=1），鲶鱼（点位 D，n=1），鳑鲏（点位 D，n=1）和沙塘鳢（点位 D，n=1）。解剖的胃内容物放置在 10%缓冲福尔马林中。在实验室中，切开胃并用解剖显微镜分析胃内容物。胃内容物被鉴定到消化状态允许的最低分类水平（赵文，2005）。通过计算每个食物来源分类单元的出现频率（$O_i\%$=包含食物的胃的数量×100/总胃数）来估算鱼类食物来源。

在本研究中，总砷及其他重金属的分析方法见第 2 章。采用水提取的方法进行砷形态分析（Caumette et al.，2011；Jia et al.，2018；Zarić et al.，2022）。据报道，水温的提高导致了更高的提取效率，而超声处理时间和离心过程对提取效率没有显著影响（Caumette et al.，2011）。碳氮同位素的分析见第 2 章。使用编程语言 R 中最新的 MixSIAR 模型建模（Stock et al.，2018），与以前的混合模型如 IsoSource、SIAR 和 Isotope R 相比，MixSIAR

模型的优势在于,它能够合并协变量数据,解释混合比例的可变性。MixSIAR 模型基于以下基本混合方程:

$$X_{ij} = \sum_{k=1}^{n} p_{ik} S_{kj} + \varepsilon_{ij}$$

式中,

X_{ij} 为第 i 个样本的第 j 个指标的值;

n 是源的数量;

p_{ik} 是第 k 个源对第 i 个样本的比例贡献;

S_{kj} 是第 k 个源中第 j 个指标的值;

ε_{ij} 是第 i 个样本的第 j 个指标的误差(He et al.,2022)。

关于 MixSIAR 模型的更多细节见文献[Stock et al.,2018]。在本研究中,利用淡水鱼的食物—肌肉辨别因子($\Delta^{13}C$ =0.47±1.23‰ 和 $\Delta^{15}N$ = 3.23±0.41‰)来估计营养分级(Zanden and Rasmussen,2001)。

应用生物放大因子(BMF)来评估重金属从食物来源向消费者的营养转移。BMF 的计算公式如下:

$$BMF = C_{水生生物} / C_{食物来源}$$

$$C_{食物来源} = \sum (c_i \times p_i)$$

式中,

$C_{水生生物}$(mg/kg)为水生生物肌肉组织中重金属的浓度;

$C_{食物来源}$(mg/kg)为水生生物所有食物来源中重金属的平均浓度;

c_i(mg/kg)为食物来源 i 中重金属的浓度;

p_i(%)为食物来源 i 在水生生物总食物来源中的平均比例。

6.3 水生生物及其食物来源对重金属的富集和砷形态分布

6.3.1 水生生物胃内容物和碳氮同位素分析

本研究利用胃内容物和碳氮同位素分析法研究水生生物中食物来源组成。但是，胃内容物分析提供的食物组成信息有限。本研究中，43个胃中有18.6%的胃都是空的，特别是点位C和点位D的泥鳅的胃也是空的（见表6.2）。

水生生物中的食物来源主要分为鱼类、腹足类、昆虫幼虫、浮游动物、浮游植物和有机残渣。根据O_i%值，来自点位A~C的马口鱼（L）中最重要的食物组成是昆虫幼虫，其次是鱼类和浮游动物。胃内容物分析表明，从石门雄黄矿采集的马口鱼（L）的食物组成主要为鱼类、昆虫幼虫、腹足类、浮游动物和浮游植物。在鲶鱼的样本中，胃内容物中发现其主要食物来源为鱼类和浮游植物。在鳑鲏的胃内容物中，发现其食物来源主要为昆虫幼虫、浮游动物、浮游植物和有机残渣。而在沙塘鳢的样品中，其胃内容物中只发现了有机残渣。

表6.2 石门雄黄矿采样鱼类胃内容物分析结果

Results of stomach content analysis of the fishes sampled in Shimen realgar mine

	马口鱼				泥鳅		鲶鱼	鳑鲏	沙塘鳢
	点位 A	点位 B	点位 C	点位 D	点位 C	点位 D	点位 D	点位 D	点位 D
胃的总数	8	8	11	11	1	1	1	1	1
空胃数量	1	2	3		1	1			
食物出现频率（O_i%）									
鱼类	+(42.86%)	+(28.57%)	+(37.50%)	+(100%)			+		
腹足类			+(12.50%)	+(36.36%)					
昆虫幼虫	+(85.71%)	+(71.43%)	+(62.50%)	+(36.36%)				+	
浮游动物	+(42.86%)	+(14.29%)	+(12.50%)	+(27.27%)				+	
浮游植物			+(12.50%)	+(18.18%)			+	+	
有机残渣								+	+

注："+"代表胃内容物中发现有此表食物来源。

石门雄黄矿黄水溪采集的水生生物及其潜在食物来源的碳氮同位素值均有所不同（见图6.1、表6.1）。水生生物的$\delta^{13}C$值范围为-25.2‰～-13.18‰，水生生物的$\delta^{15}N$值范围为7.34‰～13.79‰。点位D的青虾的$\delta^{13}C$值和$\delta^{15}N$值要远低于点位A和点位B的。点位A和点位B的马口鱼（S）的$\delta^{13}C$值和$\delta^{15}N$值要高于更大尺寸的马口鱼，但是在点位C和点位D的没有这种趋势。水生生物的潜在食物来源的$\delta^{13}C$值范围为-29.78‰～-17.31‰，水生生物的$\delta^{15}N$值范围为1.35‰～8.04‰。除点位D外，不同样品的$\delta^{13}C$值的顺序为岩石附生相>细颗粒相>粗颗粒相>枯枝落叶，而$\delta^{15}N$值的顺序为细颗粒相>岩石附生相>粗颗粒相>枯枝落叶。

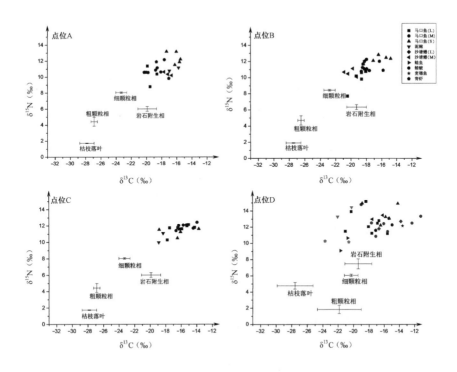

图 6.1 不同点位水生生物和潜在食物来源的 $\delta^{13}C$ 和 $\delta^{15}N$

Site-specific $\delta^{13}C$ and $\delta^{15}N$ signatures of aquatic organisms and the potential food sources

胃内容物分析对确定水生生物的食物来源具有重要作用。但是，研究需要水生生物的多种食物来源的比例，我们使用 MixSIAR 模型计算 5 种潜在食物来源：细颗粒相、岩石附生相、粗颗粒相、枯枝落叶、小鱼（本研究中采用马口鱼）。点位 A 的水生生物的主要食物来源是细颗粒相，其比例为 35%~68%，而其他点位的水生生物的主要食物来源为岩石附生相（见图 6.2）。不同尺寸的马口鱼的主要食物来源组成并无明显差异，小的马口鱼的食物组成主要为高比例的岩石附生相和低比例的细颗粒相。点位 D 的鲇鱼比其他的鱼类的食物组成中含有更高比例的小鱼，比例为 15%。此外，点位 B 的水生生物中的食物来源组成中小鱼的比例为 13%~29%。

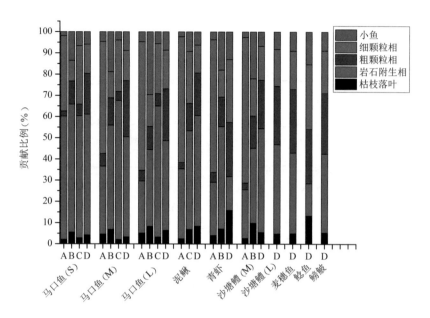

图 6.2 利用 MixSIAR 模型定量评估水生生物中每个潜在食物来源的贡献

Contribution of each potential prey items in aquatic organisms by using MixSIAR model

6.3.2 水生生物及其食物来源的重金属含量

各研究点位地表水的物理化学性质和重金属浓度如表 6.3 所示，总的来说，点位 C 和点位 D 的 As、Cd 含量远高于点位 A 和点位 B 的。TDS 和 EC 呈现从上游（点位 A 和点位 B）到下游（点位 C 和点位 D）显著的上升趋势。采集的水样均为碱性。除点位 C 和点位 D 外，采集的水样中的重金属含量均低于国家地表水环境质量标准中的要求。

第6章 食物来源影响水生生物富集和转化重金属的差异性

表 6.3 各研究点位地表水的物理化学性质和重金属浓度

Physic-chemical properties and metal (loid) concentrations of stream water at each study site

	点位 A	点位 B	点位 C	点位 D
距离（m）	−1000	+200	+1500	+2500
水温（℃）	23.1±0.03	24.4±0.05	23.1±0.01	23.1±0.02
pH 值	8.11±0.24	7.81±0.31	7.90±0.13	7.83±0.07
TDS（mg/L）	91.50±9.19	91.50±6.36	144.50±0.71	143.50±0.71
EC（μs/cm）	141.50±13.44	141.00±9.90	223.00±1.41	222.00±1.41
As（μg/L）	10.38±0.23	37.23±0.05	1389±23	1329±7
Cr（μg/L）	0.29±0.04	0.22±0.03	0.25±0.02	0.28±0.09
Cd（μg/L）	0.003±0.001	0.004±0.003	0.011±0.003	0.017±0.006
Cu（μg/L）	0.36±0.23	0.53±0.03	0.77±0.09	0.62±0.01
Mn（μg/L）	0.06±0.01	0.15±0.13	0.02±0.02	0.14±0.10
Ni（μg/L）	0.69±0.01	0.63±0.02	0.97±0.02	0.94±0.02
Pb（μg/L）	0.002±0.001	0.003±0.001	0.006±0.005	0.007±0.002
Zn（μg/L）	1.06±0.12	1.48±0.24	4.68±0.80	1.04±0.17

石门雄黄矿采集的水生生物及其潜在食物来源的重金属含量相差 2~3 个数量级左右（见图 6.3），水生生物中 As、Ni、Mn、Cu、Cr、Pb、Cd 和 Zn 的含量范围分别为 0.91~1298 mg/kg、0.31~542 mg/kg、0.82~1718 mg/kg、1.22~410 mg/kg、0.63~1158 mg/kg、0.12~37.79 mg/kg、0.01~1.30 mg/kg 和 21.84~1413 mg/kg。水生生物的潜在食物来源（粗颗粒相、细颗粒相、岩石附生相）的重金属含量最高，而水生生物的肌肉中的重金属含量相对较低。在水生生物的潜在食物来源的重金属含量中，Mn 的含量最高，Cd 的含量最低。不同样品在不同采样点间的重金属浓度存在较大差异。对于粗颗粒相，河流上游（点位 A 和点位 B）的 As、Cr、Ni、Zn 的浓度远低于下游点位 C 和点位 D 的浓度，而 Pb 和 Mn 的浓度则呈现相反的趋势。在水生生物中，来自不同采样点位的不同物种的重金属浓度有很大的差异。有趣的是，来自 3 个采样点位（点位 A、点位 B 和点位 D）的青虾的肌肉中 Cu、Cr、Cd 和 Pb 浓度高。对于马口鱼，小尺寸的马口鱼中的 Zn、Ni、Mn、Cd 和 As 浓度高于大尺寸的马口鱼。

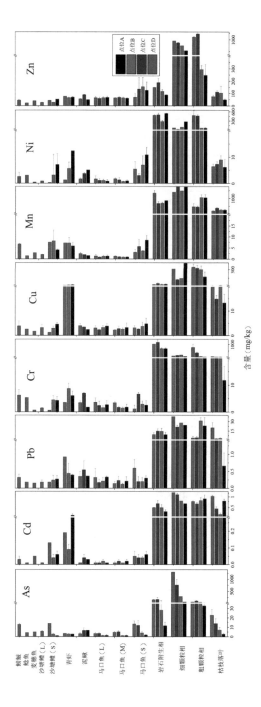

图 6.3 不同采样点位样品的重金属浓度

Concentrations of heavy metals in different samples collected at each sample site

与水生生物中的重金属含量相比，其潜在食物来源中的重金属含量更高。潜在食物来源中的 As、Cd、Pb、Cr、Cu、Mn、Ni 和 Zn 含量分别为 15.79～862 mg/kg、0.43～0.71 mg/kg、10.57～16.51 mg/kg、411～969 mg/kg、55.24～137.53 mg/kg、732～1228 mg/kg、187～450 mg/kg 和 133～458 mg/kg。除少数例外，点位 A 食物来源中 As、Cd、Pb、Cr 和 Zn 的重金属浓度低。在点位 D，鲶鱼的食物来源中 Cd、Cu 和 Zn 的浓度高，而点位 A 的马口鱼（S）食物来源中 As、Cd、Pb、Ni 和 Zn 的浓度低。除 Ni 外，不同采样点位的水生生物的肌肉中重金属浓度没有显著差异。石门雄黄矿区水生生物中重金属的生物放大因子除 As 和 Zn 外，各采样点位间无显著差异。As、Cd、Pb、Cr、Cu、Mn、Ni 和 Zn 的 BMFs 分别为 0.001～0.37、0.01～0.66、0.007～00.06、0.001～0.0.02、0.01～0.35、0.001～0.01、0.001～0.04 和 0.05～0.95（见图 6.4）。结果显示，BMFs 均<1，表明重金属从食物来源到鱼类存在一定的生物稀释作用，其顺序为 Zn>Cd>Cu>As>Pb>Ni>Cr>Mn。

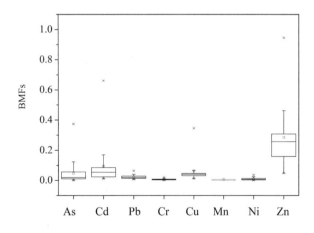

图 6.4 水生生物中重金属的生物放大因子（BMFs）

Biomagnification factor (BMFs) of heavy metal in aquatic organisms

各采样点位水生生物及其潜在食物来源的重金属浓度及其生物放大因子如表 6.4 所示。

表6.4 各采样点位水生生物及其潜在食物来源的重金属浓度及其生物放大因子

The metal(loid) concentration of aquatic organisms and its potential sources and biomagnification factors in each sampling site

		As	Cd	Pb	Cr	Cu	Mn	Ni	Zn
$C_{食物来源}$	点位A	18.35±1.51	0.45±0.02	11.34±0.46	470.3±51.76	101.3±13.82	915.1±19.61	387.34±40.56	163.41±17.65
	点位B	747.2±82	0.52±0.005	16.13±0.35	491.7±52.93	66.51±7.78	766.7±20.21	207.33±20.44	219.8±25.81
	点位C	226.8±15.96	0.67±0.01	13.88±0.43	879.4±68.04	73.59±3.36	764.39±26.70	386.15±26.99	332.57±29.42
	点位D	309.7±34.36	0.64±0.03	13.22±0.90	738.3±101.6	97.92±17.12	1124±103.3	356.29±43.86	334.9±52.8
$C_{水生生物}$	点位A	2.38±2.18	0.08±0.11	0.31±0.06	3.17±1.67	8.26±11.10	3.67±3.02	6.17±4.85	70.40±28.39
	点位B	2.31±1.15	0.04±0.03	0.240.12	3.81±2.98	7.56±10.02	4.19±3.32	3.50±2.73	74.59±47.46
	点位C	6.72±3.67	0.03±0.02	0.28±0.17	4.41±2.74	2.76±0.46	2.33±2.24	2.28±1.25	86.99±34.39
	点位D	6.91±4.55	0.05±0.06	0.33±0.25	2.93±1.93	4.57±5.84	3.56±2.54	1.89±1.49	49.90±18.12
BMFs	点位A	0.13±0.12	0.18±0.24	0.02±0.005	0.007±0.004	0.07±0.08	0.004±0.003	0.016±0.013	0.44±0.25
	点位B	0.003±0.001	0.07±0.06	0.01±0.007	0.008±0.007	0.11±0.13	0.005±0.005	0.017±0.012	0.36±0.29
	点位C	0.03±0.02	0.04±0.02	0.02±0.01	0.005±0.003	0.04±0.006	0.003±0.002	0.006±0.004	0.27±0.13
	点位D	0.02±0.01	0.08±0.09	0.03±0.02	0.004±0.003	0.04±0.08	0.003±0.003	0.005±0.01	0.15±0.06

6.3.3 水生生物及其食物来源的砷形态分布特征

我们通过HPLC-ICP-MS研究发现了6种砷形态,包括无机砷As(Ⅲ)和As(Ⅴ)、DMA、MMA和两种未知砷形态,但未检测到AsB和AsC(见表6.5)。在大多数潜在食物来源(凋落叶、粗颗粒相、细颗粒相和岩石附生相)中,主要砷形态为无机砷As(Ⅲ)和As(Ⅴ),最高可占总砷的29%和66%。在一些潜在的食物来源样本中检测到MMA、DMA和少量未知的砷1。在水生生物中,不同大小的马口鱼的砷形态组成相似,主要的砷形态为DMA、未知砷2、As(Ⅲ)和As(Ⅴ)。不同水生生物中的砷形态组成有所不同。青虾、泥鳅和鲶鱼中只有As(Ⅴ)、DMA和未知砷1,麦穗鱼中只有As(Ⅲ)、DMA和未知砷1。未知砷1在鳑鲏中最常见,而沙塘鳢和鲶鱼中的砷形态主要为DMA。然而,在本研究中,砷形态的提取率和回收率分别为7%~89%和51%~112%,这表明有大量的砷未能被提取出来,部分砷形态不能通过HPLC-ICP-MS鉴定。

第6章 食物来源影响水生生物富集和转化重金属的差异性

表 6.5 石门雄黄矿水生生物及其食物来源中的砷形态

HPLC-ICP-MS data: As compounds identified and extraction efficiency of aquatic organisms and food sources collected from the realgar mine area, China

采样点位	样品名称	砷含量 (mg/kg)	AsC	AsB	未知砷1①	As(III)	DMA	未知砷2①	MMA	As(V)	提取率 (%)	柱回收率 (%)
点位A	马口鱼(L)	1.26±0.85	nd	nd	5.96~26.20	nd	nd~9.62	nd~1.41	nd	nd~1.70	16~37	62~88
点位A	马口鱼(M)	1.17±0.53	nd	nd	16.41~49.40	nd	nd	nd~6.24	nd	nd~8.46	23~49	77~102
点位A	马口鱼(S)	1.77±0.45	nd	nd	4.10~12.35	nd	nd~7.50	nd~5.51	nd	nd~3.71	10~20	93~112
点位A	泥鳅	6.68	nd	nd	20.94	nd	3.95	nd	nd	nd	25	73
点位A	青虾	2.53±0.25	nd	nd	25.14	nd	nd	nd	nd	1.34	26	78
点位A	沙塘鳢(M)	0.91±0.26	nd	nd	nd	nd	5.39	nd	nd	18.67	24	71
点位A	岩石附生相	11.53±3.82	nd	nd	nd	nd~15.33	4.85~9.49	nd	nd	9.37~12.36	16~19	74~80
点位A	粗颗粒相	32.66±4.80	nd	nd	nd	nd	nd	nd	nd	14.97~17.94	18~30	66~81
点位A	细颗粒相	54.35	nd	nd	nd	nd	nd	nd	nd	22.02	22	97
点位B	枯枝落叶	2.70±1.16	nd	nd	nd~4.01	nd~21.09	nd~1.11	nd	nd~3.66	28.81~50.33	28~80	62~112
点位B	马口鱼(L)	1.24±0.26	nd	nd	7.07~40.39	nd~6.64	nd~15.69	nd~8.61	nd	nd	18~44	66~90
点位B	马口鱼(M)	1.12±0.22	nd	nd	9.15~22.41	nd~15.33	4.08~18.22	nd~4.21	nd~4.03	1.62~14.63	19~58	80~109
点位B	马口鱼(S)	3.91±1.14	nd	nd	10.67~36.80	nd	nd~30.03	nd	nd	nd~14.85	22~81	64~112
点位B	青虾	2.66±0.59	nd	nd	8.26~38.29	nd	nd~8.26	nd	nd	2.61~34.70	21~55	71~89
点位B	沙塘鳢(M)	2.60±0.84	nd	nd	12.12~14.88	nd~9.63	9.63~12.12	nd	nd	nd~20.12	24~54	65~91

续表

采样点位	样品名称	砷含量(mg/kg)	AsC	AsB	未知砷①	As(III)	DMA	未知砷②	MMA	As(V)	提取率(%)	柱回收率(%)
点位B	岩石附生相	1028±88.77	nd	nd	0.02~0.16	nd~21.58	nd~0.05	nd	nd~0.06	7.03~8.62	7~30	70~101
点位B	粗颗粒相	46.85±7.39	nd	nd	nd	nd~9.18	nd	nd	nd	22.69	32	72
点位B	细颗粒相	283.15	nd	nd	nd	nd	nd	nd	nd	14.85	15	96
点位B	枯枝落叶	7.10±2.80	nd	nd	nd~7.61	nd~15.27	nd~1.34	nd	nd~2.51	23.24~64.12	30~64	82~100
点位C	马口鱼(L)	3.44±1.13	nd	nd	12.84~31.46	nd~18.76	nd~17.97	nd~4.09	nd~6.56	nd~8.82	15~80	82~101
点位C	马口鱼(M)	5.02±1.24	nd	nd	12.02~41.51	nd~18.87	9.92~35.08	nd~3.46	nd	nd~8.46	30~76	62~107
点位C	马口鱼(S)	11.91±5.11	nd	nd	15.43~24.48	nd~5.99	nd~35.23	nd~3.94	nd	nd~13.77	34~60	59~108
点位C	泥鳅	6.52±2.70	nd	nd	13.61~20.49	nd	5.18~11.96	nd	nd	1.28~2.35	26~28	62~82
点位C	岩石附生相	161.15±80.02	nd	nd	nd~0.28	nd~7.36	nd~0.35	nd	nd~0.08	9.26~21.12	10~21	67~88
点位C	粗颗粒相	97.59±22.00	nd	nd	nd~0.15	nd~29.37	nd~1.27	nd	nd~1.03	20.82~50.73	22~59	63~94
点位C	细颗粒相	752.44±29.21	nd	nd	0.07	nd	0.1	nd	nd	29.16	29.85	109
点位C	枯枝落叶	13.88±5.19	nd	nd	nd~0.76	nd~22.02	nd~0.98	nd	nd~1.63	32.97~59.82	32~77	73~93
点位D	马口鱼(L)	3.28±1.27	nd	nd	3.83~27.52	nd~16.68	5.48~33.19	nd~5.79	nd~4.99	nd~14.22	16~66	69~107
点位D	马口鱼(M)	4.54±0.83	nd	nd	9.92~30.62	nd~7.78	4.29~16.12	nd~3.86	nd	nd~0.18	20~43	63~110
点位D	马口鱼(S)	13.47±3.12	nd	nd	10.94~24.52	nd~5.38	12.19~30.35	nd	nd	nd~3.89	28~51	60~106
点位D	麦穗鱼	4.87	nd	nd	23.93	12.46	6.56	nd	nd	nd	43	77
点位D	泥鳅	3.08±1.02	nd	nd	3.50~13.43	nd	10.36~16.06	nd	nd~4.56	nd	23~24	76~85
点位D	鲶鱼	4.14	nd	nd	nd~9.86	nd	36.02~44.28	nd	nd	16.27~23.10	52~77	52~109

第6章 食物来源影响水生生物富集和转化重金属的差异性

续表

采样点位	样品名称	砷含量 (mg/kg)	AsC	AsB	未知砷 1[①]	As(III)	提取砷形态比例 (%) DMA	未知砷 2[②]	MMA	As(V)	提取率 (%)	柱回收率 (%)
点位 D	鳑鲏	12.90±1.07	nd	nd	21.37~28.22	nd	4.84~8.47	nd~4.20	nd	4.79~6.10	31~47	73~86
点位 D	青虾	3.37	nd	nd	14.47	nd	nd	nd	nd	2.28	17	68
点位 D	沙塘鳢（L）	5.58±0.65	nd	nd	nd~15.95	nd	24.44~26.81	nd	nd~2.08	nd~0.04	24~42	85~90
点位 D	沙塘鳢（M）	13.87	nd	nd	7.72	nd	9.9	nd	nd~1.79	21.89	39	98
点位 D	岩石附生相	147.60±17.95	nd	nd	nd~0.27	nd~16.58	nd~0.24	nd	nd~2.08	4.90~22.29	22~24	73~87
点位 D	粗颗粒相	54.94±4.52	nd	nd	nd	nd~16.02	nd	nd	nd~1.79	44.45~65.90	40~47	91~95
点位 D	细颗粒相	1298.64	nd	nd	0.04	nd	0.04	nd	nd	13.66	13.96	84
点位 D	枯枝落叶	22.77±5.90	nd	nd	nd~0.75	nd~18.58	nd~1.09	nd	nd~2.06	30.44~66.35	30~89	51~96

注：①通过与 AsB 和 DMA 峰面积的比较，分别估算了未知砷形态的浓度。②提取效率 (%) =提取砷浓度/总砷浓度×100%。nd: 未检测到（低于检测限）或质谱仪分辨出一个峰，但低于检测限。

6.4 水生生物及其食物来源对重金属的富集和砷的转化

6.4.1 水生生物食物来源组成特征

本研究在石门雄黄矿区黄水溪采集的水生生物中共鉴定出 7 个品种,结果显示马口鱼为优势种。根据摄食习性,黄水溪的水生生物的食性主要为肉食性和杂食性。马口鱼是一种小型的杂食性鱼类,部分为肉食性,以新鲜鱼类为食物来源。小的马口鱼主要以小的猎物为食,如浮游生物和昆虫幼虫,而大的马口鱼可以消耗小虾和鱼类(Guan et al., 2022)。胃内容物分析提供的食物来源信息十分有限。根据出现频率指数,马口鱼(L)最重要的食物来源是鱼类,其次是昆虫幼虫和浮游动物。观察到的不同地点的食物来源组成差异可能与水生生物可获得的食物来源的多寡有关。

第6章 食物来源影响水生生物富集和转化重金属的差异性

本研究使用稳定同位素（$\delta^{13}C$ 和 $\delta^{15}N$）来量化食物网结构，$\delta^{15}N$ 可用来反映营养级位置，$\delta^{13}C$ 则用来反映碳源（Zanden and Rasmussen, 2001；Ai et al., 2019）。水生生物的碳氮同位素比值范围很广（其中，$\delta^{13}C$ 的差值范围达到 12.0‰，$\delta^{15}N$ 的差值范围达到 6.5‰），这与以往对淡水食物网中鱼类的碳氮同位素的研究结果一致（Marchese et al., 2014；Saigo et al., 2015；Wang et al., 2012）。在本研究中，水生生物的 $\delta^{15}N$ 值高于其潜在的食物来源，表明水生生物的营养级水平高于其食物来源。根据文献报道，捕食者的 $\delta^{15}N$ 值比其食物来源高约 3.4‰（Zanden and Rasmussen, 2001；Post, 2002），因此我们可以得出黄水溪的食物网结构至少有 3 个营养级。本研究还发现，枯枝落叶和粗颗粒相及细颗粒相等陆地碳源的 $\delta^{13}C$ 值相对较低。本书第 5 章已经显示黄水溪中浮游植物的丰度远低于中国太湖、巢湖和滇池的报道（Yang et al., 2016）。粗颗粒相和细颗粒相含有重要的有机物质和无机物质（Schubert et al., 2012）。此外，水生生物每次捕猎食物时会受到水流和天气条件的影响。由于这些原因，不同生物在不同点位的 $\delta^{13}C$ 值差异较大说明水生生物的食物来源相对丰富。

在本研究中，MixSIAR 模型的结果表明，岩石附生相和细颗粒相是水生生物的两种主要食物来源。结果显示，马口鱼所摄入的饮食与其体形大小有关。Tsai 等人（2015）利用 SIAR 模型发现，不同生长阶段的旗鱼的食物来源的消耗量存在显著差异。Li 等人（2015）也报道，当 *Leptobotia elongate* 的体长增长时，食物来源也会发生变化。更多的研究报道，食物来源组成的变化在鱼类早期生长阶段很常见，这通常是形态限制、迁移或许多鱼类栖息地变化的结果（Scharf et al., 2000；Zhang and Wang, 2007；Li et al., 2015；Lourenço et al., 2017；Riaz et al., 2020）。我们的研究结果表明，胃内容物分析和基于碳氮同位素的 MixSIAR 模型相结合更适合研究生物的食物来源组成。

水质的变化、食物来源及可获得性栖息地的丰富程度可能在很大程度上是水生生物食物来源存在差异的原因。以往的研究表明,食物来源的分布、丰度和行为的季节性变化,都会导致鱼类的食物来源可以全年发生变化（Li et al.，2015；Lourenço et al.，2017）。值得注意的是,我们于2016年6月收集的水生生物的多样性要高于本研究。本研究的样本收集仅限于9月,因此关于食物来源随季节变化的信息很少。此外,我们很难收集到足够小的野生水生生物及其潜在食物来源。基于这些原因,我们只能评估可采集到的样本,这限制了MixSIAR模型对理解这些淡水生物的自然食物来源组成的有效性。

6.4.2 水生生物及其食物来源对重金属的生物富集和营养传递

重金属一旦进入水环境中,可被颗粒相吸收、絮凝和沉淀,因此颗粒相是重金属存在的一个主要形式（Li et al.，2020）。从本研究中可以明显看出,粗颗粒相、细颗粒相和岩石附生相中的As、Cd、Pb、Cu、Cr、Mn、Ni和Zn浓度要显著高于鱼类和枯枝落叶中的重金属含量。这一结果与其他研究金赞溪大型无脊椎动物及其潜在食物来源中重金属的结果一致（Watanabe et al.，2008）。这可能是因为粗颗粒相、细颗粒相和岩石附生相的表面在运输和沉降期间可以吸收大量的重金属。因此,悬浮颗粒物可能是沉积物的替代品（Karichkoff，1984；Chamier et al.，1989；Schubert et al.，2012）。

石门雄黄矿黄水溪采集的鱼类的肌肉中的重金属浓度高于中国其他淡水生态系统的相关报道（Wei et al.，2014；Jia et al.，2017；Fang et al.，2019；Fazio et al.，2019；Yu et al.，2020；Jiang et al.，2022）。在底栖生物沙塘鳢和泥鳅中,重金属的浓度较高。在每个采样点,底栖生物青虾中的重金属总浓度普遍高于鱼类,这与之前的研究结果相似（Pereira et al.，2010；Yu et al.，2020；Basu et al.，2021）。据报道,生物体内的重金属浓度会随着生长阶段而呈现生物稀释趋势，特别对于小鱼和生长迅猛的鱼类（Hill and Larsen，2005）。在本研究中，对于马口鱼，重金属浓度随着鱼的生长而降低，这与之前的研究结果一致（Watanabe et al.，2008；Cott et al.，2016；Liu et al.，2018）。因此，我们的研究结果表明，体型大小是影响鱼类体内重金属营养转移的重要因素。

食物来源被认为是重金属富集的主要来源（>95%），而水相传播对水生生物重金属的吸收作用较少（Zhang and Wang，2007；Creighton and Twining，2010）。重金属从食物来源到生物体的营养迁移显示，从食物向捕食者的重金属浓度急剧降低。我们的研究结果表明，在水生生物中，重金属的生物富集有一种适应策略的可能性，如有效排泄或减少同化，以应对重金属毒性（Cui et al.，2021）。然而，通过 MixSIAR 模型计算出的食物来源的重金属浓度与水生生物中的重金属浓度之间没有相关性。8 种重金属（As、Cd、Pb、Cr、Cu、Mn、Ni、Zn）的 BMFs 在所有水生生物中均<1，表明具有生物稀释的潜力。这些结果与其他研究的结果一致，即重金属的积累随着营养水平的增加而减少（Culioli et al.，2009；Pereira et al.，2010；Revenga et al.，2012；Vizzini et al.，2013）。然而，Sun 等人（2020）发现，基于捕食者—食物来源的比较，As 和 Ni 呈现生物稀释的趋势，而 Pb 和 Zn 在全球海洋食物网中呈现生物放大趋势。据报道，重金属硫蛋白是确定重金属暴露的生化效应的有用工具，因为它可以防止结合和储备重金属的毒性，有效隔离它们与其他蛋白质的相互作用（Amiard et al.，2006；McDonald et al.，2021）。重金属硫蛋白诱导和重金属在细胞水平上的积累机制有待进一步研究。

6.4.3　水生生物及其食物来源对砷的转化特征

在本研究中，水生生物及其潜在食物来源中的砷形态分布呈现多样性和复杂性。DMA 是大多数鱼类中主要的砷形态，与其他已发表的淡水系统研究结果相似（Šlejkovec et al.，2004；Miyashita et al.，2009；Yang et al.，2020）。尽管 As(V)、DMA 和未鉴定的砷形态组成相似，但随着马口鱼体型的增大，As(Ⅲ)和 MMA 的比例降低。青虾中的砷形态组成与以往研究中报道的其他虾种不同（Williams et al.，2009；Hong et al.，2014；Yang et al.，2020）。Biancarosa 等人（2019）发现，虽然食物来源的总砷浓度相似，但食物来源中不同砷形态组成会影响砷在鱼类中的富集。主要食物来源细颗粒相和岩石附生相中的砷形态主要为无机砷，仅含有少量的有机砷。一种可能的解释是，浮游植物从周围的水相中吸收无机 As(V)，通过还原作用和甲基化作用产生有机砷形态（Caumette et al.，2011；Azizur Rahman et al.，2012）。据报道，微生物也具有将水生系统中的无机砷生转化为甲基砷和其他有机砷的能力（Azizur Rahman et al.，2012）。此外，本研究检测到的未知砷形态很可能是水生生物的代谢产物。但是，在文献中没有关于悬浮颗粒物中砷形态组成的相关报道。

淡水鱼中甲基砷形态（DMA 和 MMA）的比例远远高于潜在的食物来源，表明鱼类具有将无机砷转化为有机砷的能力，从而达到解毒作用（Sun et al.，2016；Cui et al.，2021）。然而，以往的实验室研究主要集中在海洋鱼类，淡水生物食物来源暴露中砷的生物富集和转化尚未得到更多研究。Kuroiwa 等人（1994）发现，通过给 *Oryzuas latipes* 喂食 *Neocaridina denticulate*，*Oryzuas latipes* 体内的砷形态主要为甲基砷，而 TMA 为唯一检测到的甲基砷形态。在其他研究中，Maeda 等人（1992，1993）发现，*Poecilia reticulata* 和 *Cyprinus carpio* 也存在同样的情况，其中，TMA 是主要的甲基化砷形态。Suhendrayatna 等人（2001，2002a）发现，喂食 *Neocaridina denticulata* 和 *Daphnia magna* 的鱼类 *Tilapia mossambica* 和 *Zacco platypus* 的砷形态主要为无机砷，仅含有少量的甲基砷。值得注意的是，*Tilapia mossambica* 体内的甲基砷主要形式是 TMA。这些结果表明，高营养水平的生物具有较强的甲基化能力。相反,也有报道发现,喂食 *D. magna* 的鱼类 *Oryzuas latipes* 中仅检测到无机砷,并未发现任何甲基砷形态(Suhendrayatna et al.,2002b)。Cui 等人(2021)研究发现,食物相 As(V)暴露后,*Carassius auratus* 具有将无机砷还原为 As(III)的高潜力。因此,对野外环境的深入研究对于建立淡水食物链中砷的生物富集和生物转化的综合模型具有重要意义。

6.5 本章小结

食物相暴露是水生生物重金属富集的主要途径,但是重金属从食物到水生生物的传递途径还不太清楚。目前的研究主要集中在室内模拟研究,本研究主要研究自然水体下重金属从食物到水生生物的传递特征。通过采集石门雄黄矿水生生物及其潜在食物来源,包括枯枝落叶、粗颗粒相、细颗粒相、附生岩石相及小鱼等,胃内容物分析和 MixSIAR 模型分析两种方法相结合,定性定量分析水生生物的食物组成及其贡献率。结果显示,附生岩石相和细颗粒相是黄水溪水生生物最主要的食物来源。不同的水生生物及其食物来源的重金属含量不同。对于淡水鱼类而言,体长和体重是重金属富集的重要影响因素。不同点位和大小的水生生物的 $\delta^{13}C$ 值和 $\delta^{15}N$ 值不同,也说明水生生物在生长过程中会出现一定的食性转变。重金属的 BMFs<1,说明水生生物对重金属具有自我调节和潜在的生物稀释作用。水生生物中的砷形态主要为有机砷,而潜在食物来源中的砷形态以无机砷为主,说明水生生物可将无机砷转化为有机砷。但是需要进一步的实验研究,确定自然条件下水生生物的食物来源组成,以阐明淡水食物网中重金属的营养转移机制。

参 考 文 献

刘其根, 吴杰洋, 颜克涛, 等. 淀山湖光泽黄颡鱼食性研究[J]. 水产学报, 2015, 39:859-866.

王玉玉, 于秀波, 张亮, 等. 应用碳、氮稳定同位素研究鄱阳湖枯水末期水生食物网结构[J]. 生态学报, 2009, 29:1181-1188.

韦丽丽, 周琼, 谢从新, 等. 三峡库区重金属的生物富集、生物放大及其生物因子的影响[J]. 环境科学, 2016, 37:325-334.

徐军. 应用碳、氮稳定性同位素探讨淡水湖泊的食物网结构和营养级关系[D]. 中国科学院水生生物研究所, 2005.

余杨, 王雨春, 周怀东, 等. 三峡水库蓄水初期鱼体汞含量及其水生食物链累积特征[J]. 生态学报, 2013, 33:4059-4067.

张金梅. 草鱼鱼种的食性[J]. 淡水渔业, 1982, 6:41-42.

张楠, 韦朝阳, 杨林生. 淡水湖泊系统中砷的赋存与转化行为研究进展[J]. 生态学报, 2013, 33:337-347.

赵文. 水生生物学[M]. 北京：中国农业出版社, 2005.

AI, S., YANG, Y., DING, J., YANG, W., BAI, X., BAO, X., JI, W, ZHANG, Y. Metal exposure risk assessment for tree sparrows at different life stages via diet from a polluted area in northwestern China[J]. Environ. Toxicol. Chem, 2019, 38: 2785-2796.

ALAMDAR, A., EQANI, S.A.M.A.S., HANIF, N., ALI, S.M., FASOLA, M., BOKHARI, H., KATSOYIANNIS, I.A., SHEN, H. Human exposure to trace metalsandarsenic via consumption of fish from river Chenab, Pakistanandassociated health risks[J]. Chemosphere, 2017, 168, 1004-1012.

AMIARD, J. C., TRIQUET, C. A., BARKA, S., PELLERIN, J.ANDRAINBOW, P. S. Metallothioneins in aquatic invertebrates: Their role in metal detoxificationandtheir use as biomarkers[J]. Aquat. Toxicol, 2006, 76: 160-202.

AZIZUR RAHMAN, M., HASEGAWA, H., LIM, R.P. Bioaccumulation, biotransformationandtrophic transfer of arsenic in the aquatic food chain[J]. Environ. Res, 2012, 116, 118-135.

BAILEY, S. E., OLIN, T. J., BRICKA, R. M., ADRIAN, D. D. A review of potentially low-cost sorbents for heavy metals[J]. Water Res, 1999, 33: 2469-2479.

BASU, S., CHANDA, A., GOGOI, P., BHATTACHARYYA, S. Organochlorine pesticidesandheavy metals in the zooplankton, fishes,andshrimps of tropical shallow tidal creeksandthe associated human health risk[J]. Mar. Pollut. Bull, 2021, 165: 112170-112184.

BEAUDOIN CP, TONN WM, PREPAS EE, et al. Individuals specializationandtrophic adaptability of northern pike (*Esox lucius*):An isotopeanddietary analysis[J]. Oecologia, 1999, 120:386-396.

BIANCAROSA, I., SELE, V., BELGHIT, I., ØRNSRUD, R., LOCK, E. J., AMLUND, H. Replacing fish meal with insect meal in the diet of Atlantic salmon (*Salmo salar)* does not impact the amount of contaminants in the feedandit lowers accumulation of arsenic in the fillet[J]. Food Addit Contam, 2019, Part A 36: 1191-1205.

BURGER J, GAINES KF, BORING S, et al. Metal levels in fish from the Savannah River: Potential hazards to fishandother receptors[J]. Environ Res, 2002, 89:95-97.

CAUMETTE, G., KOCH, I., ESTRADA, E., REIMER, K.J. Arsenic speciation in plankton organisms from contaminated lakes: transformations at the base of the freshwater food chain[J]. Environ. Sci. Technol, 2011, 45, 9917-9923.

CAUMETTE G, KOCH I, MORIARTY M, et al. Arsenic distribution and speciation in *Daphnia pulex*[J]. Sci Total Environ, 2012, 432:243-250.

CHAMIER, A. C., SUTCLIFFE, D. W., LISHMAN, J. P. Changes in Na, K, Ca, MgandAl content of submersed leaf litter, related to ingestion by the amphipod *Gammarus pulex* (L.)[J]. Freshwater Biol, 1989, 21: 181-189.

CHEN CY, FOLT CL. Bioaccumulationanddiminution of arsenicandlead in a freshwater food web [J]. Environ Sci Technol, 2000, 34:3878-3884.

CIARDULLO S, AURELI F, RAGGI A, et al. Arsenic speciation in freshwater fish: Focus on extractionandmass balance[J]. Talanta, 2010, 81:213-221.

COTT, P.A., ZAJDLIK, B.A., PALMER, M.J., MCPHERSON, M.D. Arsenicandmercury in lake whitefishandburbot near the abandoned Giant Mine on Great Slave Lake[J]. Great. Lakes Res, 2016, 42, 223-232.

CREIGHTON, N., TWINING, J. Bioaccumulation from foodandwater of cadmium, seleniumandzinc in an estuarine fish, *Ambassis jacksoniensis*[J]. Mar. Pollut. Bull, 2010, 60: 1815-1821.

CUI B, ZHANG Q, ZHANG K, et al. Analyzing trophic transfer of heavy metals for food webs in the newly-formed wetlands of the Yellow River Delta, China[J]. Environ Pollut, 2011, 159:1297-1306.

CUI, D., ZHANG, P., LI, H., ZHANG, Z., SONG, Y., YANG, Z. The dynamic changes of arsenic biotransformationandbioaccumulation in muscle of freshwater food fish crucian carp during chronic dietborne exposure[J]. Environ. Sci, 2021, 100: 74-81.

CULIOLI, J.L., FOUQUOIRE, A., CALENDINI, S., MORI, C., ORSINI, A. Trophic transfer of arsenicandantimony in a freshwater ecosystem: A field study[J]. Aquat. Toxicol, 2009, 94, 286-293.

DE ROSEMOND, S., XIE, Q., LIBER, K. Arsenic concentrationandspeciation in five freshwater fish species from Back Bay near Yellowknife, NT, Canada[J]. Environ. Monit. Assess, 2008, 147, 199-210.

DHANEESH, K.V., NOUSHAD, K.M., KUMAR, T.T.A. Nutritional evaluation of commercially important fish species of Lakshadweep Archipelago, India[J]. PLos One 7, 2012, e45439.

DOVICK, M.A., KULP, T.R., ARKLE, R.S., PILLIOD, D.S. Bioaccumulation trends of arsenicandantimony in a freshwater ecosystem affected by mine drainage[J]. Environ. Chem, 2016, 13, 149-159.

ERICKSON, R.J., MOUNT, D.R., HIGHLAND, T.L., HOCKETT, J.R., JENSON, C.T. The relative importance of waterborneanddietborne arsenic exposure on survivalandgrowth of juvenile rainbow trout[J]. Aquat. Toxicol, 2011, 104,108-115.

ERICKSON, R.J., MOUNT, D.R., HIGHLAND, T.L., HOCKETT, J.R., HOFF, D.J., JENSON, C.T., LAHREN, T.J. The effects of arsenic speciation on accumulationandtoxicity of dietborne arsenic exposures to rainbow trout[J]. Aquat. Toxicol, 2019, 210, 227-241.

FANG, T., LU, W., CUI, K., LI, J., YANG, K., ZHAO, X., LIANG, Y., LI, H. Distribution, bioaccumulationandtrophic transfer of trace metals in the food web of Chaohu Lake, Anhui, China[J]. Chemosphere, 2019, 218: 1122-1130.

FAZIO, F., SAOCA, C., FERRANTELLI, V., CAMMILLERI, G., CAPILLO G., PICCIONE G. Relationship between arsenic accumulation in tissuesandhematological parameters in mullet caught in Faro Lake: a preliminary study[J]. Environ. Sci. Pollut. R, 2019, 26:8821-8827.

FOUST JR RD, BAUER AM, COSTANZA-ROBINSON M, et al. Arsenic transferandbiotransformation in a fully characterized freshwater food web[J]. Coordin Chem Rev, 2016, 306:558-565.

FOWLER SW, ÜNLÜ MY. Factors affecting bioaccumulationandelimination of arsenic in the shrimp *Lysmata seticaudata*[J]. Chemosphere, 1978, 7:711-720.

GENNER MJ, HAWKINS SJ, TURNER GF. Isotope change throughout the life history of a Lake Malawi cichlid fish[J]. J Fish Bio, 2003, 62:907-917.

GHOSH A, MAJUMDER S, AWAL MA, et al. Arsenic exposure to dairy cows in Bangladesh[J]. Arch Environ Con Tox, 2013, 64:151-159.

GUAN, W., XU, X., ZHAN, W., NIU, B., LOU, B. Induction of gynogenesis with ultra-violet irradiated Koi carp (Cyprinus carpio haematopterus) sperm demonstrates the XX/XY sex determination system in *Opsariichthys bidens*[J]. Aquacult. Rep, 2022, 26: 101286.

HE, S., LI, P., SU, F., WANG, D., REN, X. Identificationandapportionment of shallow groundwater nitrate pollution in Weining Plain, northwest China, using hydrochemical indices, nitrate stable isotopes,andthe new Bayesian stable isotope mixing model (MixSIAR)[J]. Environ. Pollut, 2022, 298: 118852.

HILL, W. R., LARSEN, I. L. Growth dilution of metals in microalgal biofilms[J]. Environ. Sci. Technol, 2005, 39: 1513-1518.

HONG S, KHIM JS, PARK J, et al. Species-andtissue-specific bioaccumulation of arsenicals in various aquatic organisms from a highly industrialized area in the

Pohang City, Korea[J]. Environ Pollut, 2014, 192:27-35.

HUGHES MF. Arsenic toxicityandpotential mechanisms of action[J]. Toxicol Lett, 2002, 133:1-16.

JIA, Y., WANG, L., QU, Z., WANG, H., YANG, Z. Effects on heavy metal accumulation in freshwater fishes: species, tissues,andsizes[J]. Environ. Sci. Pollut. R, 2017, 24: 9379-9386.

JIA, Y.Y., WANG, L., LI, S., CAO, J.F., YANG, Z.G. Species-specific bioaccumulationandcorrelated health risk of arsenic compounds in freshwater fish from a typical mine-impacted river[J]. Sci. Total Environ, 2018, 625, 600-607.

JIANG, X., WANG, J., PAN, B., LI, D., WANG, Y., LIU, X. Assessment of heavy metal accumulation in freshwater fish of Dongting Lake, China: Effects of feeding habits, habitat preferencesandbody size[J]. Environ. Sci, 2022, 112: 355-365.

KARICHKOFF, S. W. Organic pollutant sorption in aquatic systems[J]. Hydraul. Eng. ASCE, 1984, 110: 707-735.

KUROIWA T, OHKI A, NAKA K, et al. Biomethylation and biotransformation of arsenic in a freshwater food chain: Green alga (*Chlorella vulgaris*)→shrimp (*Neocaridina denticulata*)→killifish (*Oryzias iatipes*)[J]. Appl Organometal Chem, 1994, 8:325-333.

LI, L., WEI, Q. W., WU, J. M., ZHANG, H., LIU, Y., XIE, X. Diet of *Leptotia elongata* revealed by stomach content analysisandinferred from stable isotope signatures[J]. Environ. Biol. Fishes, 2015, 98: 1965-1978.

LI, R., TANG, X., GUO, W., LIN, L., ZHAO, L., HU, Y., LIU, M. Spatiotemporal distribution dynamics of heavy metals in water, sediment, andzoobenthos in mainstream sections of middleandlower Changjiang River[J].

Total Environ, 2020, 20: 136779.

LI, Z., MA, Z., VAN DER KUIJP, T. J., YUAN, Z., HUANG, L. A review of soil heavy metal pollution from mines in China: Pollutionandhealth risk assessment[J]. Sci Total Environ, 2014, 468-469: 843-853.

LIU Y, LIU G, YUAN Z, et al. Heavy metals (As, Hg and V) and stable isotope ratios (δ^{13}C and δ^{15}N) in fish from Yellow River Estuary, China[J]. Sci Total Environ, 2018, 613-614:462-471.

LOURENÇO, S., A.SAUNDERS, R., COLLINS, M., SHREEVE, R., ASSIS, C. A., BELCHIER, M., WATKINS, J. L., XAVIER, J. C. Life cycle, distributionandtrophodynamics of the lanternfish Krefftichthys anderssoni (Lnnberg, 1905) in the Scotia Sea[J]. Polar Biol, 2017, 40: 1229-1245.

MAEDA, S., OHKI, A., KUSADOME, K., KUROIWA, T., YOSHIFUKU, I., NAKA, K. Bioaccumulation of arsenicandits fate in a freshwater food chain[J]. Appl. Organomet. Chem, 1992, 6: 213-219.

MAEDA S, MAWATARI K, OHKI A, et al. Arsenic metabolism in a freshwater food chain: Blue–green alga (*Nostoc sp.*)→shrimp (*Neocaridina denticulata*)→carp (*Cyprinus carpio*)[J]. Appl Organometal Chem, 1993, 7:467-476.

MANDAL, B.K., SUZUKI, K.T. Arsenic round the world: a review[J]. Talanta, 2002, 58, 201-235.

MARCHESE MR, SAIGO M, ZILLI FL, et al. Food webs of the Paraná River floodplain: assessing basal sources using stable carbon and nitrogen isotopes[J]. Limnologica, 2014, 46:22-30.

MATHEWS, T., FISHER, N. S. Dominance of dietary intake of metals in marine elasmobranchandteleost fish[J]. Total Environ, 2009, 407: 5156-5161.

MCDONALD, S., HASSELL, K., CRESSWELL, T. Effect of short-term dietary exposure on metal assimilationandmetallothionein induction in the estuarine fish *Pseudogobius* sp[J]. Total Environ, 2021, 772: 145042.

MIYASHITA S, SHIMOYA M, KAMIDATE Y, et al. Rapid determination of arsenic species in freshwater organisms from the arsenic-rich Hayakawa River in Japan using HPLC-ICP-MS[J]. Chemosphere, 2009, 75:1065-1073.

MOORE JW, SUTHERLAND DJ, BEAUBIEN VA, et al. Algalandinvertebrate communities in three subarctic lakes receiving mine wastes[J]. Water Res, 1979, 13:1193-1202.

MOORE, J. W., SEMMENS, B. X. Incorporating uncertaintyandprior information into stable isotope mixing models[J]. Ecol. Lett, 2008, 11: 470-480.

MORI C, ORSINI A, MIGON C. Impact of arsenicandantimony contamination on benthic invertebrates in a minor Corsican river[J]. Hydrobiologia, 1999, 392:73-80.

O'MARA, K., ADAMS, M., BURFORD, M. A., FRY, B., CRESSWELL, T. Uptakeandaccumulation of cadmium, manganeseandzinc by fisheries species: Trophic differences in sensitivity to environmnetal metal accumulation[J]. Sci Total Environ, 2019, 690: 867-877.

OVERMAN NC, PARRISH DL. Stable isotope composition of walleye: 15N accumulation with age and area-specific differences in $\delta^{13}C$[J]. Can J Fish Aquat Sci, 2001, 58:1253-1260.

PARNELL, A. C., INGER, R., BEARHOP, S., JACKSON, A. L. Source partitioning using stable isotopes: coping with too much variation[J]. Plos ONE, 2010, 5: e9672.

PEDLAR, R.M., KLAVERKAMP, J.F. Accumulationanddistribution of dietary arsenic in lake whitefish (*Coregonus clupeaformis*)[J]. Aquat. Toxicol, 2002, 57, 153-166.

PEREIRA, A.A., VAN HATTUM, B., DE BOER, J., VAN BODEGOM, P.M., REZENDE, C.E., SALOMONS, W. Trace elementsandcarbonandnitrogen stable isotopes in organisms from a tropical coastal lagoon[J]. Arch. Environ. Contam. Toxicol, 2010, 59, 464-477.

PHILIPS, D. L., GREGG, J. W. Source partitioning using stable isotopes: coping with too many sources[J]. Oecologia, 2003, 136: 261-269.

PHILLIPS DL, NEWSOME SD, GREGG JW. Combining sources in stable isotope mixing models: alternative methods[J]. Oecologia, 2005,144:520-527.

POST, D.M. Using stable isotopes to estimate trophic position: Models, methods,andassumptions[J]. Ecology, 2002, 83, 703-718.

POUIL, S., BUSTAMANTE, P., WARNAU, M., METIAN, M. Overview of trace elements trophic transfer in fish through the concept of assimilation efficiency[J]. Mar. Ecol. Prog. Ser, 2018, 588: 243-254.

REVENGA, J.E., CAMPBELL, L.M., ARRIBÉRE, M.A., GUEVARA, S.R. Arsenic, cobaltandchromium food web biodilution in a Patagonia mountain lake[J]. Ecotox. Environ. Safe, 2012, 81, 1-10.

RIAZ, J., WALTERS, A., TREBILCO, R., BESTLEY, S., LEA, M. A. Stomach content analysis of mesopelagic fish from the southern Kerguelen Axis[J]. Deep-Sea Res. Part II, 2020, 174: 104659.

SAIGO, M., ZILLI, F.L., MARCHESE, M.R., DEMONTE, D. Trophic level, food chain lengthandomnivory in the Paraná River: a food web model approach in a floodplain river system[J]. Ecol. Res, 2015, 30, 843-852.

SCHAEFFER, R., FRANCESCONI, K.A., KIENZL, N., SOEROES, C.,

FODOR, P., VÁRADI, L., RAML, R., GOESSLER, W., KUEHNELT, D. Arsenic speciation in freshwater organisms from the river Danube in Hungary[J]. Talanta, 2006, 69, 856-865.

SCHARF, F. S., JUANES, F.ANDROUNTREE, R. A. Predator size prey size relationships of marine fish predators: interspecific variationandeffects of ontogenyandbody size trophic-niche breadth[J]. Mar. Ecol. Prog. Ser, 2000, 208: 229-248.

SCHUBERT, B., HEININGER, P., KELLER, M., RICKING, M., CLAUS, E. Monitoring of contaminants in suspended particulate matter as an alternative to sediments[J]. Trend. Anal. Chem, 2012, 36: 58-70.

SHEN S, LI XF, CULLEN WR, et al. Arsenic binding to proteins[J]. Chem Rev, 2013, 113:7769-7792.

ŠLEJKOVEC, Z., BAJC, Z., DOGANOC, D.Z. Arsenic speciation patterns in freshwater fish[J]. Talanta, 2004, 62, 931-936.

SOEROES C, GOESSLER W, FRANCESCONI KA, et al. Arsenic speciation in farmed Hungarian freshwater fish [J]. J Agric Food Chem, 2005, 53:9238-9243.

STOCK, B. C., JACKSON, A. L., WARD, E. J., PARNELL, A. C., PHILLIPS, D. L.ANDSEMMENS, B. X. Analyzing mixing system using a new generation of Bayesian tracer mixing models[J]. Peer J Preprints, 2018, 6:e26884v1.

SUHENDRAYATNA, OHKI, A., MAEDA, S. Biotransformation of arsenite in freshwater food chain models[J]. Appl. Organometal. Chem, 2001, 15, 277-284.

SUHENDRAYATNA, OHKI, A., NAKAJIMA, T., MAEDA, S. Studies on

the accumulationandtransformation of arsenic in freshwater organisms II. Accumulationandtransformation of arsenic compounds by *Tilapia mossambica*[J]. Chemosphere, 2002, 46: 325-331.

SUHENDRAYATNA, OHKI, A., NAKAJIMA, T.ANDMAEDA, S. Studies on the accumulationandtransformation of arsenic in freshwater organisms I. Accumulation, transformationandtoxicity of arsenic compounds on the Japanese Medaka, *Oryzias latipes*[J]. Chemosphere, 2002, 46: 319-324.

SUN, T., WU, H., WANG, X., JI, C., SHAN, X., LI, F. Evaluation on the biomagnification or biodilution of trace metals in global marine food webs by meta-analysis[J]. Environ. Pollut, 2020, 264: 113856.

SUN, Y., LIU, G., CAI, Y. Thiolated arsenicals in arsenic metabolism: Occurrence, formation,andbiological implications[J]. Environ Sci, 2016, 49:59-73.

TANG, J.W., LIAO, Y.P., YANG, Z.H., CHAI, L.Y., YANG, W.C. Characterization of arsenic seriouscontaminated soils from Shimen realgar mine area, the Asian largest realgar deposit in China[J]. Soil Sediment, 2016, 16,1519-1528.

TSAI, C. N., CHIANG, W. C., SUN, C. L., SHAO, K. T., CHEN, S. Y., YEH, S. Z. Stomach contentandstable isotope analysis of sailfish (*Istiophorus platypterus*) diet in eastern Taiwan waters[J]. Fish. Res, 2015, 166: 39-46.

VANDER ZANDEN, M.J., RASMUSSEN, J.B. Primary consumer $\delta^{13}C$ and $\delta^{15}N$ and the trophic position of aquatic consumers[J]. Ecology, 1999, 80, 1395-1404.

VANDER ZANDEN, M. J., RASMUSSEN, J.B. Variation of $\delta^{13}C$ and $\delta^{15}N$ trophic fractionation: Implication for aquatic food web studies[J]. Limnol. Oceanogr, 2001, 46: 2061-2066.

VARELA, J. L., SORELL, J. M., LAIZ-CARRIÓN, R., BARO, I., URIARTE, A., MACÍAS, D., MEDINA, A. Stomach contentandstable isotope analyses reveal resource partitioning between juvenile bluefin tunaandAtlantic bonito in Alboran (SW Mediterranean)[J]. Fish. Res, 2019, 215: 97-105.

VENTURA-LIMA J, RAMOS PB, FATTORINI D, et al. Accumulation, biotransformation, and biochemical responses after exposure to arseniteandarsenate in the estuarine polychaete *Laeonereis acuta* (Nereididae)[J]. Environ Sci Pollut Res, 2011, 18:1270-1278.

VIZZINI, S., COSTA, V., TRAMATI, C., GIANGUZZA, P., MAZZOLA, A. Trophic transfer of trace elements in an isotopically constructed food chain from a semi-enclosed marine coastal area (Stagnone di Marsala, Sicily, Mediterranean)[J]. Arch. Environ. Contam. Tox, 2013, 65, 642-653.

WANG, W., FISHER, N. S. Assimilation efficiencies of chemical contaminants in aquatic invertebrates: A synthesis[J]. Environ. Toxicol. Chem, 1999, 18: 2034-2045.

WANG S, LI B, ZHANG M, et al. Bioaccumulation and trophic transfer of mercury in a food web from a large, shallow, hypereutrophic lake (Lake Taihu) in China[J]. Environ Sci Pollut Res, 2012, 19:2820-2831.

WATANABE, K., MONAGHAN, M.T., TAKEMON, Y., OMURA, T. Biodilution of heavy metals in a stream macroinvertebrate food web: Evidence from stable isotope analysis[J]. Sci Total Environ, 2008, 394, 57-67.

WEI, B., YANG, L. A review of heavy metals contaminations in urban soils, urban road dustsandagricultural soils from China[J]. Microchem, 2010, 94: 99-107.

WEI, Y., ZHANG, J., ZHANG, D., TU, T., LUO, L. Metal concentrations in various fish organs of dfifferent fish species from Poyang Lake, China. Ecotoxicol[J]. Environ. Saf, 2014, 104: 182-188.

WILLIAMS, G., WEST, J.M., KOCH, I., REIMER, K.J., SNOW, E.T. Arsenic speciation in the freshwater crayfish, *Cherax destructor* Clark[J]. Sci Total Environ, 2009, 407, 2650-2658.

WILLIAMS, J.J., DUTTON, J., CHEN, C.Y., FISHER, N.S. Metal (As, Cd, Hg, and CH_3Hg) bioaccumulation from waterandfood by the benthic amphipod *Leptocheirus plumulosus*[J]. Environ. Toxicol. Chem, 2010, 29, 1755-1761.

XIAO, W., LIN, G., HE, X., YANG, Z., WANG, L. Interactions among heavy metal bioaccessibility, soil propertiesandmicrobial community in phyto-remediated soils nearby an abandoned realgar mine[J]. Chemosphere, 2022, 286: 131638.

YANG, F., GENG, D., WEI, C. Distribution of arsenic between the particulateandaqueous phases in surface water from three freshwater lakes in China. Environ[J]. Sci Pollut. R, 2016, 23: 7452-7461.

YANG, F., YU, Z., XIE, S., FENG, H., WEI, C., ZHANG, H., ZHANG, J. Application of stable isotopes to the bioaccumulationandtrophic transfer of arsenic in aquatic organisms around a closed realgar mine[J]. Sci Total Environ, 2020, 726: 138550-138562.

YU, B., WANG, X., DONG, K. F., XIAO, G., MA, D. Heavy metal concentrations in aquatic organisms (fishes, shrimpandcrabs) andhealth risk assessment in China[J]. Mar. Pollut. Bull, 2020, 159: 111505-111512.

ZARIĆ, N. M., BRAEUER, S., GOESSLER, W. Arsenic speciation analysis in honey bees for environmental monitoring[J]. Hazard. Mater, 2022, 432: 128614-128619.

ZHANG, L., WANG, W. Size-dependence of the potential for metal biomagnification in early life stage of marine fish[J]. Environ. Toxicol. Chem, 2007, 26: 787-794.

ZHU X, WANG R, LU X, et al. Secondary minerals of weathered orpiment-realgar-bearing tailings in Shimen carbonate-type realgar mine, Changde, Central China[J]. Miner Petrol, 2015,109:1-15.

ZHU, Y.M., WEI, C.Y., YANG, L.S. Rehabilitation of a tailing dam at Shimen County, Hunan Province: Effectiveness assessment[J]. Acta Ecologica Sinica, 2010, 30, 178-183.